# Lecture Notes on Mathematical Olympiad Courses

### For Junior Section Vol. 2

# Mathematical Olympiad Series

ISSN: 1793-8570

**Series Editors:** Lee Peng Yee *(Nanyang Technological University, Singapore)*
Xiong Bin *(East China Normal University, China)*

*Published*

Xu Jiagu

**Vol. 6** | Mathematical
Olympiad
Series

# Lecture Notes on Mathematical Olympiad Courses

For Junior Section Vol. 2

**World Scientific**

*Published by*

World Scientific Publishing Co. Pte. Ltd.

5 Toh Tuck Link, Singapore 596224

*USA office:* 27 Warren Street, Suite 401-402, Hackensack, NJ 07601

*UK office:* 57 Shelton Street, Covent Garden, London WC2H 9HE

**British Library Cataloguing-in-Publication Data**
A catalogue record for this book is available from the British Library.

First published 2010
Reprinted with corrections 2011

Mathematical Olympiad Series — Vol. 6
**LECTURE NOTES ON MATHEMATICAL OLYMPIAD COURSES**
**For Junior Section**

ISBN-13 978-981-4293-53-2  (pbk) (Set)
ISBN-10 981-4293-53-9      (pbk) (Set)

ISBN-13 978-981-4293-54-9  (pbk) (Vol. 1)
ISBN-10 981-4293-54-7      (pbk) (Vol. 1)

ISBN-13 978-981-4293-55-6  (pbk) (Vol. 2)
ISBN-10 981-4293-55-5      (pbk) (Vol. 2)

Printed in Singapore by World Scientific Printers.

# Preface

Although mathematical olympiad competitions are carried out by solving problems, the system of Mathematical Olympiads and the related training courses cannot involve only the techniques of solving mathematical problems. Strictly speaking, it is a system of mathematical advancing education. To guide students who are interested in mathematics and have the potential to enter the world of Olympiad mathematics, so that their mathematical ability can be promoted efficiently and comprehensively, it is important to improve their mathematical thinking and technical ability in solving mathematical problems.

An excellent student should be able to think flexibly and rigorously. Here the ability to do formal logic reasoning is an important basic component. However, it is not the main one. Mathematical thinking also includes other key aspects, like starting from intuition and entering the essence of the subject, through prediction, induction, imagination, construction, design and their creative abilities. Moreover, the ability to convert concrete to the abstract and vice versa is necessary.

Technical ability in solving mathematical problems does not only involve producing accurate and skilled-computations and proofs, the standard methods available, but also the more unconventional, creative techniques.

It is clear that the usual syllabus in mathematical educations cannot satisfy the above requirements, hence the mathematical olympiad training books must be self-contained basically.

The book is based on the lecture notes used by the editor in the last 15 years for Olympiad training courses in several schools in Singapore, like Victoria Junior college, Hwa Chong Institution, Nanyang Girls High School and Dunman High School. Its scope and depth significantly exceeds that of the usual syllabus, and introduces many concepts and methods of modern mathematics.

The core of each lecture are the concepts, theories and methods of solving mathematical problems. Examples are then used to explain and enrich the lectures, and indicate their applications. And from that, a number of questions are included for the reader to try. Detailed solutions are provided in the book.

The examples given are not very complicated so that the readers can understand them more easily. However, the practice questions include many from actual competitions which students can use to test themselves. These are taken from a range of countries, e.g. China, Russia, the USA and Singapore. In particular, there are many questions from China for those who wish to better understand mathematical Olympiads there. The questions are divided into two parts. Those in Part A are for students to practise, while those in Part B test students' ability to apply their knowledge in solving real competition questions.

Each volume can be used for training courses of several weeks with a few hours per week. The test questions are not considered part of the lectures, since students can complete them on their own.

# Acknowledgments

My thanks to Professor Lee Peng Yee for suggesting the publication of this the book and to Professor Phua Kok Khoo for his strong support. I would also like to thank my friends, Ang Lai Chiang, Rong Yifei and Gwee Hwee Ngee, lecturers at HwaChong, Tan Chik Leng at NYGH, and Zhang Ji, the editor at WSPC for her careful reading of my manuscript, and their helpful suggestions. This book would be not published today without their efficient assistance.

# Abbreviations and Notations

## Abbreviations

ix

USAMO    United States of American Mathematical Olympiad
USSR    Union of Soviet Socialist Republics

# Notations for Numbers, Sets and Logic Relations

| | |
|---|---|
| $\mathbb{N}$ | the set of positive integers (natural numbers) |
| $\mathbb{N}_0$ | the set of non-negative integers |
| $\mathbb{Z}$ | the set of integers |
| $\mathbb{Z}^+$ | the set of positive integers |
| $\mathbb{Q}$ | the set of rational numbers |
| $\mathbb{Q}^+$ | the set of positive rational numbers |
| $\mathbb{Q}_0^+$ | the set of non-negative rational numbers |
| $\mathbb{R}$ | the set of real numbers |
| $[m, n]$ | the lowest common multiple of the integers $m$ and $n$ |
| $(m, n)$ | the greatest common devisor of the integers $m$ and $n$ |
| $a \mid b$ | $a$ divides $b$ |
| $\mid x \mid$ | absolute value of $x$ |
| $\lfloor x \rfloor$ | the greatest integer not greater than $x$ |
| $\lceil x \rceil$ | the least integer not less than $x$ |
| $\{x\}$ | the decimal part of $x$, i.e. $\{x\} = x - \lfloor x \rfloor$ |
| $a \equiv b \pmod{c}$ | $a$ is congruent to $b$ modulo $c$ |
| $\binom{n}{k}$ | the binomial coefficient $n$ choose $k$ |
| $n!$ | $n$ factorial, equal to the product $1 \cdot 2 \cdot 3 \cdot n$ |
| $[a, b]$ | the closed interval, i.e. all $x$ such that $a \leq x \leq b$ |
| $(a, b)$ | the open interval, i.e. all $x$ such that $a < x < b$ |
| $\Leftrightarrow$ | iff, if and only if |
| $\Rightarrow$ | implies |
| $A \subset B$ | $A$ is a subset of $B$ |
| $A - B$ | the set formed by all the elements in $A$ but not in $B$ |
| $A \cup B$ | the union of the sets $A$ and $B$ |
| $A \cap B$ | the intersection of the sets $A$ and $B$ |
| $a \in A$ | the element $a$ belongs to the set $A$ |

# Contents

# Lecture 16

# Quadratic Surd Expressions and Their Operations

## Definitions

For an even positive integer $n$, by the notation $\sqrt[n]{a}$, where $a \geq 0$, we denote the non-negative real number $x$ which satisfies the equation $x^n = a$. In particular, when $n = 2$, $\sqrt[2]{a}$ is called **square root of $a$**, and denoted by $\sqrt{a}$ usually.

For odd positive integer $n$ and any real number $a$, by the notation $\sqrt[n]{a}$ we denote the real number $x$ which satisfies the equation $x^n = a$.

An algebraic expression containing $\sqrt{a}$, where $a > 0$ is not a perfect square number, is called **quadratic surd expression**, like $1 - \sqrt{2}$, $\dfrac{1}{2 - \sqrt{3}}$, etc.

## Basic Operational Rules on $\sqrt{a}$

(I)    $(\sqrt{a})^2 = a$, where $a \geq 0$.

(II)   $\sqrt{a^2} = |a| = \begin{cases} a & \text{for } a > 0, \\ 0 & \text{for } a = 0, \\ -a & \text{for } a < 0. \end{cases}$

(III)  $\sqrt{ab} = \sqrt{|a|} \cdot \sqrt{|b|}$    if $ab \geq 0$.

(IV)   $\sqrt{\dfrac{a}{b}} = \dfrac{\sqrt{|a|}}{\sqrt{|b|}}$    if $ab \geq 0$, $b \neq 0$.

(V)    $(\sqrt{a})^n = \sqrt{a^n}$    if $a \geq 0$.

(VI)   $a\sqrt{c} + b\sqrt{c} = (a + b)\sqrt{c}$    if $c \geq 0$.

1

**Rationalization of Denominators**

(I)    $\dfrac{1}{a\sqrt{b}+c\sqrt{d}} = \dfrac{a\sqrt{b}-c\sqrt{d}}{a^2b-c^2d}$, where $a,b,c,d$ are rational numbers, $b,d \geq 0$ and $a^2b-c^2d \neq 0$.

(II)   $\dfrac{1}{a\sqrt{b}-c\sqrt{d}} = \dfrac{a\sqrt{b}+c\sqrt{d}}{a^2b-c^2d}$, where $a,b,c,d$ are rational numbers, $b,d \geq 0$ and $a^2b-c^2d \neq 0$.

In algebra, the expressions $A + B\sqrt{C}$ and $A - B\sqrt{C}$, where $A, B, C$ are rational and $\sqrt{C}$ is irrational, are called **conjugate surd expressions**.

The investigation of surd forms is necessary and very important in algebra, since surd forms and irrational number have close relation. For example, all the numbers of the form $\sqrt{n}, n \in \mathbb{N}$ are irrational if the positive integer $n$ is not a perfect square. In other words, the investigation of surd form expressions is the investigation of irrational numbers and their operations essentially.

**Examples**

**Example 1.** Simplify the expression $\dfrac{a}{a-2b}\sqrt{\dfrac{a^2-4ab+4b^2}{a(2b-a)}}$.

**Solution**   Since $a - 2b \neq 0$, so

$$\frac{a^2-4ab+4b^2}{a(2b-a)} = \frac{(a-2b)^2}{a(2b-a)} > 0 \Rightarrow a(2b-a) > 0.$$

Therefore $\dfrac{a}{a-2b} < 0$ and $\dfrac{a}{2b-a} > 0$, so

$$\frac{a}{a-2b}\sqrt{\frac{a^2-4ab+4b^2}{a}} = -\left(\frac{a}{2b-a}\right)\sqrt{\frac{(2b-a)^2}{a(2b-a)}}$$

$$= -\sqrt{\frac{a^2}{(2b-a)^2}\cdot\frac{(2b-a)}{a}} = -\sqrt{\frac{a}{2b-a}}.$$

**Example 2.** Given that $c > 1$ and

$$x = \frac{\sqrt{c+2}-\sqrt{c+1}}{\sqrt{c}-\sqrt{c-1}}, \quad y = \frac{\sqrt{c+2}-\sqrt{c+1}}{\sqrt{c+1}-\sqrt{c}}, \quad z = \frac{\sqrt{c}-\sqrt{c-1}}{\sqrt{c+2}-\sqrt{c+1}},$$

Arrange $x, y, z$ in ascending order.

**Solution**  From

$$x = \frac{\sqrt{c+2}-\sqrt{c+1}}{\sqrt{c}-\sqrt{c-1}} = \frac{[(\sqrt{c+2})^2-(\sqrt{c+1})^2](\sqrt{c}+\sqrt{c-1})}{(\sqrt{c+2}+\sqrt{c+1})[(\sqrt{c})^2-(\sqrt{c-1})^2]}$$

$$= \frac{\sqrt{c}+\sqrt{c-1}}{\sqrt{c+2}+\sqrt{c+1}},$$

$$y = \frac{\sqrt{c+2}-\sqrt{c+1}}{\sqrt{c+1}-\sqrt{c}} = \frac{[(\sqrt{c+2})^2-(\sqrt{c+1})^2](\sqrt{c+1}+\sqrt{c})}{(\sqrt{c+2}+\sqrt{c+1})[(\sqrt{c+1})^2-(\sqrt{c})^2]}$$

$$= \frac{\sqrt{c+1}+\sqrt{c}}{\sqrt{c+2}+\sqrt{c+1}},$$

it follows that $x < y$. Further,

$$z = \frac{\sqrt{c}-\sqrt{c-1}}{\sqrt{c+2}-\sqrt{c+1}} = \frac{[(\sqrt{c})^2-(\sqrt{c-1})^2](\sqrt{c+2}+\sqrt{c+1})}{(\sqrt{c}+\sqrt{c-1})[(\sqrt{c+2})^2-(\sqrt{c+1})^2]}$$

$$= \frac{\sqrt{c+2}+\sqrt{c+1}}{\sqrt{c}+\sqrt{c-1}}.$$

Since $\sqrt{c}+\sqrt{c-1} < \sqrt{c+1}+\sqrt{c} < \sqrt{c+2}+\sqrt{c+1}$, thus $x < y < z$.

**Example 3.** (SSSMO/2003) Let $x$ be a real number, and let

$$A = \frac{-1+3x}{1+x} - \frac{\sqrt{|x|-2}+\sqrt{2-|x|}}{|2-x|}.$$

Prove that $A$ is an integer, and find the unit digit of $A^{2003}$.

**Solution**  Since $|x|-2 \geq 0$ and $2-|x| \geq 0$ simultaneously implies $|x| = 2$, so $x = \pm 2$ only. Since the denominator $|x-2| \neq 0$, i.e. $x \neq 2$, so $x = -2$. Therefore $A = 7$. Then

$$7^{2003} = (7^4)^{500} \cdot 7^3 \equiv 243 \equiv 3 \pmod{10},$$

therefore the units digit of $A$ is 3.

**Example 4.** Given $x = \dfrac{\sqrt{7}+\sqrt{3}}{\sqrt{7}-\sqrt{3}}$, $y = \dfrac{\sqrt{7}-\sqrt{3}}{\sqrt{7}+\sqrt{3}}$, find the value of $x^4 + y^4 + (x+y)^4$.

**Solution**  Here an important technique is to express to $x^4 + y^4 + (x+y)^4$ by $x+y$ and $xy$ instead of using the complicated expression of $x$ and $y$. From

$$x = \frac{1}{7-3}(\sqrt{7}+\sqrt{3})^2 = \frac{1}{4}(10+2\sqrt{21}) = \frac{1}{2}(5+\sqrt{21}),$$

$$y = \frac{1}{7-3}(\sqrt{7}-\sqrt{3})^2 = \frac{1}{4}(10-2\sqrt{21}) = \frac{1}{2}(5-\sqrt{21}),$$

it follows that $x + y = 5$ and $xy = 1$. Therefore

$$x^4 + y^4 + (x + y)^4$$
$$= (x^2 + y^2)^2 - 2x^2y^2 + 5^4 = [(x + y)^2 - 2(xy)]^2 - 2(xy)^2 + 625$$
$$= 23^2 - 2 + 625 = 527 + 625 = 1152.$$

**Example 5.** Simplify the expression $\dfrac{\sqrt{2} + \sqrt{3} - \sqrt{5}}{\sqrt{2} + \sqrt{3} + \sqrt{5}}$ by rationalizing the denominator.

**Solution**

$$\frac{\sqrt{2} + \sqrt{3} - \sqrt{5}}{\sqrt{2} + \sqrt{3} + \sqrt{5}} = 1 - \frac{2\sqrt{5}}{\sqrt{2} + \sqrt{3} + \sqrt{5}}$$
$$= 1 - \frac{2\sqrt{5}(\sqrt{2} + \sqrt{3} - \sqrt{5})}{(\sqrt{2} + \sqrt{3})^2 - 5} = 1 - \frac{2(\sqrt{10} + \sqrt{15} - 5)}{2\sqrt{2} \cdot \sqrt{3}}$$
$$= 1 - \frac{\sqrt{10} + \sqrt{15} - 5}{\sqrt{6}} = 1 - \frac{\sqrt{60} + \sqrt{90} - 5\sqrt{6}}{6}$$
$$= 1 - \frac{\sqrt{15}}{3} - \frac{\sqrt{10}}{2} + \frac{5\sqrt{6}}{6}.$$

**Example 6.** Simplify $S = \sqrt{x^2 + 2x + 1} - \sqrt{x^2 + 4x + 4} + \sqrt{x^2 - 6x + 9}$.

**Solution**   From $S = \sqrt{x^2 + 2x + 1} - \sqrt{x^2 + 4x + 4} + \sqrt{x^2 - 6x + 9} = |x + 1| - |x + 2| + |x - 3|$, there are four possible cases as follows:

(i)     When $x \le -2$, then $S = -(x + 1) + (x + 2) - (x - 3) = -x + 4$.

(ii)    When $-2 < x \le -1$, then $S = -(x + 1) - (x + 2) - (x - 3) = -3x$.

(iii)   When $-1 < x \le 3$, then $S = (x + 1) - (x + 2) - (x - 3) = -x + 2$.

(iv)    When $3 < x$, then $S = (x + 1) - (x + 2) + (x - 3) = x - 4$.

**Example 7.** (SSSMO/2002/Q12) Evaluate

$$(\sqrt{10} + \sqrt{11} + \sqrt{12})(\sqrt{10} + \sqrt{11} - \sqrt{12})(\sqrt{10} - \sqrt{11} + \sqrt{12})(\sqrt{10} - \sqrt{11} - \sqrt{12}).$$

**Solution**   Let $A = (\sqrt{10} + \sqrt{11} + \sqrt{12})(\sqrt{10} + \sqrt{11} - \sqrt{12})(\sqrt{10} - \sqrt{11} + \sqrt{12})(\sqrt{10} - \sqrt{11} - \sqrt{12})$. Then

$$A = [(\sqrt{10} + \sqrt{11})^2 - (\sqrt{12})^2][(\sqrt{10} - \sqrt{11})^2 - (\sqrt{12})^2]$$
$$= (9 + 2\sqrt{10} \cdot \sqrt{11})(9 - 2\sqrt{10} \cdot \sqrt{11}) = 81 - 440 = -359.$$

**Example 8.** Evaluate $N = \dfrac{\sqrt{15} + \sqrt{35} + \sqrt{21} + 5}{\sqrt{3} + 2\sqrt{5} + \sqrt{7}}$.

**Solution** $N = \dfrac{(\sqrt{15} + \sqrt{21}) + (\sqrt{35} + 5)}{(\sqrt{3} + \sqrt{5}) + (\sqrt{5} + \sqrt{7})} = \dfrac{(\sqrt{3} + \sqrt{5})(\sqrt{5} + \sqrt{7})}{(\sqrt{3} + \sqrt{5}) + (\sqrt{5} + \sqrt{7})}$

$\Longrightarrow \dfrac{1}{N} = \dfrac{(\sqrt{3} + \sqrt{5}) + (\sqrt{5} + \sqrt{7})}{(\sqrt{3} + \sqrt{5})(\sqrt{5} + \sqrt{7})} = \dfrac{1}{\sqrt{5} + \sqrt{7}} + \dfrac{1}{\sqrt{3} + \sqrt{5}}$

$\qquad = \dfrac{1}{2}(\sqrt{7} - \sqrt{5}) + \dfrac{1}{2}(\sqrt{5} - \sqrt{3}) = \dfrac{1}{2}(\sqrt{7} - \sqrt{3}).$

$\therefore N = \dfrac{2}{\sqrt{7} - \sqrt{3}} = \dfrac{2(\sqrt{7} + \sqrt{3})}{4} = \dfrac{\sqrt{7} + \sqrt{3}}{2}.$

**Example 9.** (Training question for National Team of Canada) Simplify

$P = \dfrac{1}{2\sqrt{1} + \sqrt{2}} + \dfrac{1}{3\sqrt{2} + 2\sqrt{3}} + \cdots + \dfrac{1}{100\sqrt{99} + 99\sqrt{100}}.$

**Solution** For each positive integer $n$,

$\dfrac{1}{(n+1)\sqrt{n} + n\sqrt{n+1}} = \dfrac{1}{\sqrt{n(n+1)}(\sqrt{n+1} + \sqrt{n})} = \dfrac{\sqrt{n+1} - \sqrt{n}}{\sqrt{n(n+1)}}$

$\qquad\qquad\qquad\qquad = \dfrac{1}{\sqrt{n}} - \dfrac{1}{\sqrt{n+1}},$

hence

$P = \left(1 - \dfrac{1}{\sqrt{2}}\right) + \left(\dfrac{1}{\sqrt{2}} - \dfrac{1}{\sqrt{3}}\right) + \cdots + \left(\dfrac{1}{\sqrt{99}} - \dfrac{1}{\sqrt{100}}\right)$

$\quad = 1 - \dfrac{1}{\sqrt{100}} = 1 - \dfrac{1}{10} = \dfrac{9}{10}.$

## Testing Questions   (A)

1. If $x < 2$, then $|\sqrt{(x-2)^2} + \sqrt{(3-x)^2}|$ is equal to

(A) $5 - 2x$     (B) $2x - 5$     (C) 2     (D) 3.

2. Simplify $\dfrac{1 + \sqrt{2} + \sqrt{3}}{1 - \sqrt{2} + \sqrt{3}}$ by rationalizing the denominator.

3. Simplify the expression $\dfrac{x^2 - 4x + 3 + (x+1)\sqrt{x^2 - 9}}{x^2 + 4x + 3 + (x-1)\sqrt{x^2 - 9}}$, where $x > 3$.

4.   Simplify $\dfrac{2 + 3\sqrt{3} + \sqrt{5}}{(2 + \sqrt{3})(2\sqrt{3} + \sqrt{5})}$.

5.   Evaluate

$$(\sqrt{5} + \sqrt{6} + \sqrt{7})(\sqrt{5} + \sqrt{6} - \sqrt{7})(\sqrt{5} - \sqrt{6} + \sqrt{7})(-\sqrt{5} + \sqrt{6} + \sqrt{7}).$$

6.   (SSSMO(J)/1999) Suppose that $a = \sqrt{6} - 2$ and $b = 2\sqrt{2} - \sqrt{6}$. Then
     (A) $a > b$,   (B) $a = b$,   (C) $a < b$,   (D) $b = \sqrt{2}a$,   (E) $a = \sqrt{2}b$.

7.   Arrange the three values $a = \sqrt{27} - \sqrt{26}, b = \sqrt{28} - \sqrt{27}, c = \sqrt{29} - \sqrt{28}$
     in ascending order

8.   The number of integers $x$ which satisfies the inequality $\dfrac{3}{1 + \sqrt{3}} < x <$
     $\dfrac{3}{\sqrt{5} - \sqrt{3}}$ is
     (A) 2,      (B) 3,      (C) 4,      (D) 5,      (E) 6.

9.   Calculate the value of $\dfrac{1}{1 - \sqrt[4]{5}} + \dfrac{1}{1 + \sqrt[4]{5}} + \dfrac{2}{1 + \sqrt{5}}$.

10.  Given $a > b > c > d > 0$, and $U = \sqrt{ab} + \sqrt{cd}$, $V = \sqrt{ac} + \sqrt{bd}$, $W = \sqrt{ad} + \sqrt{bc}$. Use "<" to connect $U, V, W$.

## Testing Questions    (B)

1.   (CHINA/1993) Find the units digit of the expression

$$x = \left( \frac{-2a}{4 + a} - \frac{\sqrt{|a| - 3} + \sqrt{3 - |a|}}{3 - a} \right)^{1993}.$$

2.   (CHNMOL/1993) Simplify $\sqrt[3]{3} \left( \sqrt[3]{\dfrac{4}{9}} - \sqrt[3]{\dfrac{2}{9}} + \sqrt[3]{\dfrac{1}{9}} \right)^{-1}$.

3.   (CHINA/1998) Evaluate $\sqrt{\dfrac{1998 \times 1999 \times 2000 \times 2001 + 1}{4}}$.

4.   Given $a = \sqrt[3]{4} + \sqrt[3]{2} + 1$, find the value of $\dfrac{3}{a} + \dfrac{3}{a^2} + \dfrac{1}{a^3}$.

5.   Given that the decimal part of $M = (\sqrt{13} + \sqrt{11})^6$ is $P$, find the value of $M(1 - P)$.

# Lecture 17

# Compound Quadratic Surd Form $\sqrt{a \pm \sqrt{b}}$

**Basic Methods for Simplifying Compound Surd Forms**

(I)    Directly simplify according to algebraic formulas: like

$$\sqrt{(a+b)^2} = |a+b|, \quad \sqrt{(a+b)^4} = (a+b)^2, \quad \sqrt[3]{(a+b)^3} = a+b, \text{ etc.}$$

(II)   Use the techniques for completing squares to change the expression inside the outermost square root sign to a square, like the simplification of $\sqrt{2+\sqrt{3}}$.

(III)  Use other methods like Coefficient-determining method, substitutions of variables, etc.

**Examples**

**Example 1.** (SSSMO(J)/2003) Find the value of $\sqrt{17 + 4\sqrt{13}} - \sqrt{17 - 4\sqrt{13}}$.

  **Solution**

$$\sqrt{17 + 4\sqrt{13}} - \sqrt{17 - 4\sqrt{13}}$$
$$= \sqrt{(\sqrt{13} + 2)^2} - \sqrt{(\sqrt{13} - 2)^2} = \sqrt{13} + 2 - (\sqrt{13} - 2) = 4.$$

**Example 2.** (SSSMO(J)/2002) Find the value of $\sqrt{\dfrac{2}{5 - 2\sqrt{6}}} - \sqrt{\dfrac{2}{5 + 2\sqrt{6}}}$.

7

**Solution**   Since $5 - 2\sqrt{6} = (\sqrt{3} - \sqrt{2})^2, 5 + 2\sqrt{6} = (\sqrt{3} + \sqrt{2})^2$,

$$
\sqrt{\frac{2}{5 - 2\sqrt{6}}} - \sqrt{\frac{2}{5 + 2\sqrt{6}}} = \sqrt{\frac{2}{(\sqrt{3} - \sqrt{2})^2}} - \sqrt{\frac{2}{(\sqrt{3} + \sqrt{2})^2}}
$$

$$
= \frac{\sqrt{2}}{\sqrt{3} - \sqrt{2}} - \frac{\sqrt{2}}{\sqrt{3} + \sqrt{2}} = \frac{\sqrt{2}(\sqrt{3} + \sqrt{2}) - \sqrt{2}(\sqrt{3} - \sqrt{2})}{(\sqrt{3})^2 - (\sqrt{2})^2}
$$

$$
= (\sqrt{6} + 2) - (\sqrt{6} - 2) = 4.
$$

**Example 3.** Simplify $\sqrt{4 + \sqrt{15}} + \sqrt{4 - \sqrt{15}} - 2\sqrt{3 - \sqrt{5}}$.

**Solution**   Since

$$
\sqrt{4 + \sqrt{15}} = \frac{1}{\sqrt{2}} \cdot \sqrt{8 + 2\sqrt{15}} = \frac{1}{\sqrt{2}} \sqrt{(\sqrt{5} + \sqrt{3})^2} = \frac{\sqrt{5} + \sqrt{3}}{\sqrt{2}},
$$

$$
\sqrt{4 - \sqrt{15}} = \frac{1}{\sqrt{2}} \cdot \sqrt{8 - 2\sqrt{15}} = \frac{1}{\sqrt{2}} \sqrt{(\sqrt{5} - \sqrt{3})^2} = \frac{\sqrt{5} - \sqrt{3}}{\sqrt{2}},
$$

$$
\sqrt{3 - \sqrt{5}} = \frac{1}{\sqrt{2}} \cdot \sqrt{6 - 2\sqrt{5}} = \frac{1}{\sqrt{2}} \sqrt{(\sqrt{5} - 1)^2} = \frac{\sqrt{5} - 1}{\sqrt{2}},
$$

so

$$
\sqrt{4 + \sqrt{15}} + \sqrt{4 - \sqrt{15}} - 2\sqrt{3 - \sqrt{5}}
$$

$$
= \frac{(\sqrt{5} + \sqrt{3}) + (\sqrt{5} - \sqrt{3}) - 2(\sqrt{5} - 1)}{\sqrt{2}} = \frac{2}{\sqrt{2}} = \sqrt{2}.
$$

**Example 4.** Simplify $M = \sqrt{2 + \sqrt{-2 + 2\sqrt{5}}} - \sqrt{2 - \sqrt{-2 + 2\sqrt{5}}}$.

**Solution**   Let $a = \sqrt{2 + \sqrt{-2 + 2\sqrt{5}}}, b = \sqrt{2 - \sqrt{-2 + 2\sqrt{5}}}$. Then $a^2 + b^2 = 4$ and

$$
ab = \sqrt{4 - (-2 + 2\sqrt{5})} = \sqrt{6 - 2\sqrt{5}} = \sqrt{5} - 1.
$$

Therefore $(a - b)^2 = 4 - 2(\sqrt{5} - 1) = 6 - 2\sqrt{5} = (\sqrt{5} - 1)^2$, so

$$
M = a - b = \sqrt{5} - 1.
$$

**Example 5.** Simplify $\sqrt{9 + 2(1 + \sqrt{3})(1 + \sqrt{7})}$.

**Solution**   Considering that $9 + 2(1 + \sqrt{3})(1 + \sqrt{7}) = 11 + 2\sqrt{3} + 2\sqrt{5} + 2\sqrt{15}$, where the coefficients of the terms of $\sqrt{3}, \sqrt{5}, \sqrt{15}$ are all 2, it is natural to use the coefficient-determining method, assume that

$$\sqrt{9 + 2(1 + \sqrt{3})(1 + \sqrt{7})} = \sqrt{a} + \sqrt{b} + \sqrt{c}.$$

Taking squares on both sides yields

$$11 + 2\sqrt{3} + 2\sqrt{5} + 2\sqrt{15} = a + b + c + 2\sqrt{ab} + 2\sqrt{ac} + 2\sqrt{bc}.$$

By the comparison of coefficients, the following system of equations is obtained:

$$
\begin{aligned}
a + b + c &= 11, & (17.1) \\
ab &= 3, & (17.2) \\
ac &= 5, & (17.3) \\
bc &= 15. & (17.4)
\end{aligned}
$$

$(17.2) \times (17.3) \times (17.4)$ yields $(abc)^2 = 15^2$, i.e. $abc = 15$, so $a = 1$ from $(17.4)$, $b = 3$ from $(17.3)$, and $c = 5$ from $(17.1)$. Thus,

$$\sqrt{9 + 2(1 + \sqrt{3})(1 + \sqrt{7})} = \sqrt{1} + \sqrt{3} + \sqrt{5}.$$

**Example 6.** (SSSMO(J)/2007) Find the value of $\dfrac{x^4 - 6x^3 - 2x^2 + 18x + 23}{x^2 - 8x + 15}$ when $x = \sqrt{19 - 8\sqrt{3}}$.

**Solution**   $x = \sqrt{19 - 8\sqrt{3}} = \sqrt{(4 - \sqrt{3})^2} = 4 - \sqrt{3}$ yields $4 - x = \sqrt{3}$. By taking squares, it follows that

$$x^2 - 8x + 13 = 0.$$

Hence, by long division,

$$x^4 - 6x^3 - 2x^2 + 18x + 23 = (x^2 - 8x + 13)(x^2 + 2x + 1) + 10 = 10,$$

so that the value of the given expression is $10/2 = 5$.

**Example 7.** Given that the integer part and fractional part of $\sqrt{37 - 20\sqrt{3}}$ are $x$ and $y$ respectively. Find the value of $x + y + \dfrac{4}{y}$.

**Solution**   $\sqrt{37 - 20\sqrt{3}} = 5 - 2\sqrt{3} = 1 + 2(2 - \sqrt{3})$ implies that

$$x = 1, \qquad \text{and} \qquad y = 2(2 - \sqrt{3}),$$

hence

$$x + y + \frac{4}{y} = 5 - 2\sqrt{3} + \frac{2}{(2 - \sqrt{3})} = 5 - 2\sqrt{3} + 2(2 + \sqrt{3}) = 9.$$

**Example 8.** Given that $y$ is the nearest integer of $\sqrt{\dfrac{2}{\sqrt[3]{3} - 1}} + \sqrt[3]{3}$, find the value

of $\sqrt{9 + 4\sqrt{y}}$.

**Solution**   Since $2 = (\sqrt[3]{3})^3 - 1 = (\sqrt[3]{3} - 1)(\sqrt[3]{9} + \sqrt[3]{3} + 1)$,

$$\sqrt{\frac{2}{\sqrt[3]{3} - 1}} + \sqrt[3]{3} = \sqrt{\sqrt[3]{9} + \sqrt[3]{3} + 1 + \sqrt[3]{3}} = \sqrt{(\sqrt[3]{3} + 1)^2} = \sqrt[3]{3} + 1.$$

It is clear that $2 < \sqrt[3]{3} + 1 < 3$. Further, $(1.5)^3 > 3 \implies 2.5 - (\sqrt[3]{3} + 1) = 1.5 - \sqrt[3]{3} > \sqrt[3]{3} - \sqrt[3]{3} = 0$, so $2 < \sqrt[3]{3} + 1 < 2.5$, hence $y = 2$. Thus

$$\sqrt{9 + 4\sqrt{y}} = \sqrt{9 + 4\sqrt{2}} = \sqrt{(\sqrt{8} + 1)^2} = 2\sqrt{2} + 1.$$

**Example 9.** Simplify $\sqrt[3]{a + \dfrac{a + 8}{3}\sqrt{\dfrac{a - 1}{3}}} + \sqrt[3]{a - \dfrac{a + 8}{3}\sqrt{\dfrac{a - 1}{3}}}.$

**Solution**   Let $x = \sqrt{\dfrac{a - 1}{3}}$, then $a = 3x^2 + 1$ and $\dfrac{a + 8}{3} = x^2 + 3$, so that the given expression can be expressed in terms of $x$:

$$\sqrt[3]{a + \frac{a + 8}{3}\sqrt{\frac{a - 1}{3}}} + \sqrt[3]{a - \frac{a + 8}{3}\sqrt{\frac{a - 1}{3}}}$$

$$= \sqrt[3]{3x^2 + 1 + (x^2 + 3)x} + \sqrt[3]{3x^2 + 1 - (x^2 + 3)x}$$

$$= \sqrt[3]{x^3 + 3x^2 + 3x + 1} + \sqrt[3]{1 - 3x + 3x^2 - x^3} = \sqrt[3]{(x + 1)^3} + \sqrt[3]{(1 - x)^3}$$

$$= (x + 1) + (1 - x) = 2.$$

**Example 10.** Find the value of $\sqrt{2\sqrt{2\sqrt{2\sqrt{2\cdots}}}} - \sqrt{2 + \sqrt{2 + \sqrt{2 + \sqrt{2 + \cdots}}}}.$

**Solution**   Let $x = \sqrt{2\sqrt{2\sqrt{2\sqrt{2\cdots}}}}$, $y = \sqrt{2 + \sqrt{2 + \sqrt{2 + \sqrt{2 + \cdots}}}}$.
Then $x$ satisfies the equation

$$x^2 = 2x,$$

and its solution is $x = 2$ (since $x > 0$). Similarly, $y$ satisfies the equation

$$y^2 = 2 + y.$$

Then $(y - 2)(y + 1) = 0$ and $y > 0$ yields the solution $y = 2$. Thus

$$\sqrt{2\sqrt{2\sqrt{2\sqrt{2\cdots}}}} - \sqrt{2 + \sqrt{2 + \sqrt{2 + \sqrt{2 + \cdots}}}} = x - y = 0.$$

## Testing Questions    (A)

1.  Simplify $\sqrt{12 - 4\sqrt{5}}$.

2.  (CHINA/1996) Simplify $\sqrt{2 + \sqrt{3}} + \sqrt{2 - \sqrt{3}}$.

3.  (CHNMOL/2000) Evaluate $\sqrt{14 + 6\sqrt{5}} - \sqrt{14 - 6\sqrt{5}}$.

4.  (CHINA/1998) Evaluate $\sqrt{8 + \sqrt{63}} - \sqrt{8 - \sqrt{63}}$.

5.  Simplify $\sqrt{4 + \sqrt{7}} + \sqrt{4 - \sqrt{7}}$.

6.  (CHINA/1994) Simplify $\sqrt{7 - \sqrt{15} - \sqrt{16 - 2\sqrt{15}}}$

7.  (CHINA/1998) Given $x + y = \sqrt{3\sqrt{5} - \sqrt{2}}, x - y = \sqrt{3\sqrt{2} - \sqrt{5}}$, find the value of $xy$.

8.  Simplify $\sqrt{8 + 2(2 + \sqrt{5})(2 + \sqrt{7})}$.

9.  Evaluate $\sqrt{a + 3 + 4\sqrt{a - 1}} + \sqrt{a + 3 - 4\sqrt{a - 1}}$.

10. Let $A = \dfrac{\sqrt{\sqrt{3} + 1} - \sqrt{\sqrt{3} - 1}}{\sqrt{\sqrt{3} + 1} + \sqrt{\sqrt{3} - 1}}$. Is $A$ the root of the equation $x + 2 = \dfrac{\sqrt{6} - \sqrt{30}}{\sqrt{2} - \sqrt{10}}$?

## Testing Questions   (B)

1. (CHINA/1996) Simplify $\sqrt{2\sqrt{ab} - a - b}$, where $a \neq b$.

2. Simplify $\sqrt[3]{\dfrac{(\sqrt{a-1} - \sqrt{a})^5}{\sqrt{a-1} + \sqrt{a}}} + \sqrt[3]{\dfrac{(\sqrt{a-1} + \sqrt{a})^5}{\sqrt{a} - \sqrt{a-1}}}$.

3. Simplify $\sqrt{1 + a^2 + \sqrt{1 + a^2 + a^4}}$.

4. Simplify $\sqrt{x + 2 + 3\sqrt{2x - 5}} - \sqrt{x - 2 + \sqrt{2x - 5}}$.

5. Given $\sqrt{x} = \sqrt{a} - \dfrac{1}{\sqrt{a}}$, find the value of $\dfrac{x + 2 + \sqrt{x^2 + 4x}}{x + 2 - \sqrt{x^2 + 4x}}$.

6. (CHINA/1999) Find the nearest integer of $\dfrac{1}{\sqrt{17 - 12\sqrt{2}}}$.

# Lecture 18

# Congruence of Integers

**Definition 1**  When an integer $n$ is divided by a non-zero integer $m$, there must be an integral quotient $q$ and a remainder $r$, where $0 \leq |r| < m$. This relation is denoted by

$$n = mq + r,$$

and the process for getting this relation is called **division with remainder**.

**Definition 2**  Two integers $a$ and $b$ are said to be **congruent modulo $m$**, denoted by $a \equiv b \pmod{m}$, if $a$ and $b$ have the same remainder when they are divided by a non-zero integer $m$. If the remainders are different, then $a$ and $b$ are said to be **not congruent modulo $m$**, denoted by $a \not\equiv b \pmod{m}$.

By the definition of congruence, the following four equivalent relations are obvious:

$$a \equiv b \pmod{m} \Longleftrightarrow a - b = km \Longleftrightarrow a - b \equiv 0 \pmod{m} \Longleftrightarrow m \mid (a - b).$$

## Basic Properties of Congruence

(I)  If $a \equiv b \pmod{m}$ and $b \equiv c \pmod{m}$, then $a \equiv c \pmod{m}$.

(II)  If $a \equiv b \pmod{m}$ and $c \equiv d \pmod{m}$, then

$$(a + c) \equiv (b + d) \pmod{m}, \quad (a - c) \equiv (b - d) \pmod{m}.$$

(III)  If $a \equiv b \pmod{m}$ and $c \equiv d \pmod{m}$, then $a \cdot c \equiv b \cdot d \pmod{m}$.

(IV)  If $a \equiv b \pmod{m}$ then $a^n \equiv b^n \pmod{m}$ for all natural numbers $n$.

(V)  If $ac \equiv bc \pmod{m}$ and $(c, m) = 1$, then $a \equiv b \pmod{m}$.

13

**The Units Digit of Powers of Positive Integers $a^n$**

Let $P$ be the units digit of a positive integer $a$, and $n$ be the positive integer power of $a$. Then the units digit of $a^n$ is determined by the units digits of $P^n$, denoted by $U(P^n)$, and the sequence $\{U(P^n),\ n = 1, 2, 3, \ldots\}$ follows the following rules:

(I)     The sequence takes constant values for $P = 0, 1, 5, 6$, i.e. $U(P^n)$ does not change as $n$ changes.

(II)    The sequence is periodic with a period 2 for $P = 4$ or 9.

(III)   The sequence is periodic with a period 4 for $P = 2, 3, 7, 8$.

**The Last Two digits of some positive integers**

(I)     The last two digits of $5^n$ ($n \geq 2$) is 25.

(II)    The ordered pair of last two digits of $6^n$ ($n \geq 2$) changes with the period "36, 16, 96, 76, 56" as $n$ changes.

(III)   The ordered pair of last two digits of $7^n$ ($n \geq 2$) changes with the period "49, 43, 01, 07" as $n$ changes.

(IV)    The ordered pair of last two digits of $76^n$ is always 76.

**Examples**

**Example 1.** (CHINA/2004) When a three digit number is divided by 2, 3, 4, 5 and 7, the remainders are all 1. Find the minimum and maximum values of such three digit numbers.

   **Solution**   Let $x$ be a three digit with the remainder 1 when divided by 2, 3, 4, 5 and 7. Then $x - 1$ is divisible by each of 2, 3, 4, 5, 7, so

$$x - 1 = k \cdot [2, 3, 4, 5, 7] = 420k.$$

Thus, the minimum value of $x$ is $420 + 1 = 421$, the maximum value of $x$ is $2 \times 420 + 1 = 841$.

**Example 2.** It is known that $2726, 4472, 5054, 6412$ have the same remainder when they are divided by some two digit natural number $m$. Find the value of $m$.

   **Solution**   For excluding the effect of the unknown remainder, the three differences formed by the four given numbers can be used to replace the original four numbers. Then

$$m \mid (4472 - 2726) \Rightarrow m \mid 1746. \qquad 1746 = 2 \cdot 3^2 \cdot 97;$$
$$m \mid (5054 - 4472) \Rightarrow m \mid 582. \qquad 582 = 2 \cdot 3 \cdot 97;$$
$$m \mid (6412 - 5054) \Rightarrow m \mid 1358. \qquad 1358 = 2 \cdot 7 \cdot 97.$$

Since 97 is the unique two digit common divisor of the differences, so $m = 97$.

**Example 3.** (CHINA/2000) Find the remainder of $3^{2000}$ when it is divided by 13.

**Solution** $3^3 = 27 \equiv 1 \pmod{13}$ provides the method for reducing the power of 3, it follows that

$$3^{2000} \equiv (3^3)^{666} \cdot 3^2 \equiv 3^2 \equiv 9 \pmod{13}.$$

Thus, the remainder is 9.

**Note:** For finding the remainder of a large power of a positive integer, it's important to find the minimum power with remainder 1, or see if the remainders are constant as the power changes.

**Example 4.** (SSSMO(J)/2001) Find the smallest positive integer $k$ such that $2^{69} + k$ is divisible by 127.

**Solution** $2^7 \equiv 1 \pmod{127}$ implies $2^{7m} \equiv (2^7)^m \equiv 1^m \equiv 1 \pmod{127}$, hence

$$2^{69} = [(2^7)^9](2^6) \equiv 2^6 \equiv 64 \pmod{127},$$

therefore the minimum value of $k$ is equal to $127 - 64 = 63$.

**Example 5.** (SSSMO/2003) What is the remainder when $6^{273} + 8^{273}$ is divided by 49?

**Solution** In general, for odd positive integer $n$,

$$a^n + b^n = (a + b)(a^{n-1} - a^{n-2}b + a^{n-3}b^2 - \cdots + b^{n-1}),$$

so that

$$6^{273} + 8^{273} = (6 + 8)(6^{272} - 6^{271} \cdot 8 + 6^{270} \cdot 8^2 - \cdots + 8^{272}) = 14M,$$

where $M = 6^{272} - 6^{271} \cdot 8 + 6^{270} \cdot 8^2 - \cdots + 8^{272}$. Furthermore,

$$M \equiv \underbrace{(-1)^{272} - (-1)^{271} + (-1)^{270} - \cdots + 1}_{273 \text{ terms}} \equiv 273 \equiv 0 \pmod{7},$$

therefore $7 \mid M$, hence $49 \mid 14M$ i.e. the remainder is 0.

**Example 6.** Find the remainder of the number $2005^{2007^{2009}}$ when divided by 7.

**Solution**   First of all $2005^{2007^{2009}} \equiv 3^{2007^{2009}}$ (mod 7). Since $3^3 \equiv -1$ (mod 7) yields $3^6 \equiv (3^3)^2 \equiv 1$ (mod 7),

$$2007^{2009} \equiv 3^{2009} \equiv 3 \quad \text{(mod 6)},$$

it follows that $2007^{2009} = 6k + 3$ for some positive integer $k$. Therefore

$$2005^{2007^{2009}} \equiv 3^{6k+3} \equiv 3^3 \equiv 6 \quad \text{(mod 7)}.$$

Thus, the remainder is 6.

**Example 7.** (SSSMO/1997) Find the smallest positive integer $n$ such that $1000 \le n \le 1100$ and $1111^n + 1222^n + 1333^n + 1444^n$ is divisible by 10.

**Solution**   Let $N = 1111^n + 1222^n + 1333^n + 1444^n$. Then

$$N \equiv 1^n + 2^n + 3^n + 4^n \quad \text{(mod 10)}.$$

For estimating the minimum value of $n$, we test $n = 1000$. Then

$$N \equiv 1 + (2^4)^{250} + (3^4)^{250} + (4^2)^{500} \equiv 1 + 6 + 1 + 6 \equiv 4 \quad \text{(mod 10)}.$$

Hence for $n = 1001$,

$$N \equiv 1 + 6 \cdot 2 + 1 \cdot 3 + 6 \cdot 4 \equiv 1 + 2 + 3 + 4 \equiv 0 \quad \text{(mod 10)}.$$

Thus $n_{\min} = 1001$.

**Example 8.** Prove that for any odd natural number $n$, the number $1^{2007} + 2^{2007} + \cdots + n^{2007}$ is not divisible by $n + 2$.

**Solution**   By taking modulo $n + 2$, and partition the terms as groups of two each, $1^{2007} + 2^{2007} + \cdots + n^{2007}$

$$= 1 + \left(2^{2007} + n^{2007}\right) + \cdots + \left(\left(\frac{n+1}{2}\right)^{2007} + \left(\frac{n+3}{2}\right)^{2007}\right)$$

$$\equiv 1 + \left(2^{2007} + (-2)^{2007}\right) + \cdots + \left(\left(\frac{n+1}{2}\right)^{2007} + \left(-\frac{n+1}{2}\right)^{2007}\right)$$

$$\equiv 1 \quad \text{(mod } n + 2\text{)}.$$

Thus, the conclusion is proven.

**Example 9.** (SSSMO(J)/2001) Write down the **last four digits** of the number $7^{128}$.

**Solution** Here the recursive method is effective. Start from $7^4 = 2401$, then

$7^4 = 2401 \equiv 2401 \pmod{10^4}$,

$7^8 = (7^4)^2 = (2400 + 1)^2 = (2400)^2 + 4800 + 1 \equiv 4801 \pmod{10^4}$,

$7^{16} \equiv (4800 + 1)^2 \equiv 9601 \pmod{10^4}$,

$7^{32} \equiv (9600 + 1)^2 \equiv 9201 \pmod{10^4}$,

$7^{64} \equiv (9200 + 1)^2 \equiv 8401 \pmod{10^4}$,

$7^{128} \equiv (8400 + 1)^2 \equiv 6801 \pmod{10^4}$.

Therefore the last four digits of $7^{128}$ is 6801.

## Testing Questions    (A)

1.  (CHINA/2001) Find the number of positive integer $n$, such that the remainder is 7 when 2007 is divided by $n$.

2.  (SSSMO/1999) What is the remainder of $123456789^4$ when it is divided by 8?

3.  Prove that $7 \mid (2222^{5555} + 5555^{2222})$.

4.  Find the remainder of $47^{37^{27}}$ when it is divided by 11.

5.  (CHINA/1990) What is the remainder when $9^{1990}$ is divided by 11?

6.  (CHINA/2004) $n = 3 \times 7 \times 11 \times 15 \times 19 \times \cdots \times 2003$. Find the last three digits of $n$.

7.  (CHINA/2002) When a positive integer $n$ is divided by 5, 7, 9, 11, the remainders are 1, 2, 3, 4 respectively. Find the minimum value of $n$.

8.  (IMO/1964) (a) Find all positive integers $n$ for which $2^n - 1$ is divisible by 7.

    (b) Prove that there is no positive integer $n$ for which $2^n + 1$ is divisible by 7.

9.  (SSSMO/00/Q11) What is the units digit of $3^{1999} \times 7^{2000} \times 17^{2001}$?

    (A) 1     (B) 3     (C) 5     (D) 7     (E) 9

10.  Find the last two digits of $2^{999}$.

## Testing Questions    (B)

1.  Find the last two digits of $14^{14^{14}}$.

2.  Find the remainder of $(257^{33} + 46)^{26}$ when it is divided by 50.

3.  (SSSMO(J)/2003) What is the smallest positive integer $n > 1$ such that $3^n$ ends with 003?

4.  (CHNMOL/1997) There is such a theorem: "If three prime numbers $a, b, c >$ 3 satisfy the relation $2a + 5b = c$, then $a + b + c$ is divisible by the integer $n$." What is the maximum value of the possible values of $n$? Prove your conclusion.

5.  (MOSCOW/1982) Find all the positive integers $n$, such that $n \cdot 2^n + 1$ is divisible by 3.

# Lecture 19

# Decimal Representation of Integers

**Definition** The **decimal representation of integers** is the number system that takes 10 as the base. Under this representation system, an $(n + 1)$-digit whole number (where $n$ is a non-negative integer) $N = \overline{a_n a_{n-1} \cdots a_1 a_0}$ means

$$N = a_n \times 10^n + a_{n-1} \times 10^{n-1} + \cdots + a_1 \times 10 + a_0. \qquad (19.1)$$

The advantage of the representation (19.1) is that a whole number is expanded as $n + 1$ independent parts, so that even though there may be unknown digits, the operations of addition, subtraction and multiplication on integers can be carried out easily.

## Decimal Expansion of Whole Numbers with Same Digits or Periodically Changing Digits

$$\underbrace{\overline{aaa \cdots a}}_{n} = a(10^{n-1} + 10^{n-2} + \cdots + 10 + 1) = \frac{a}{9}(10^n - 1),$$

$$\underbrace{\overline{abcabc \cdots abc}}_{n \text{ of } \overline{abc}} = \overline{abc}(10^{3(n-1)} + 10^{3(n-2)} + \cdots + 10^3 + 1) = \frac{\overline{abc}}{999}(10^{3n} - 1).$$

## Examples

**Example 1.** (MOSCOW/1983) Find the smallest whole number such that its first digit is 4, and the value of the number obtained by moving this 4 to the last place is $\frac{1}{4}$ of the original value.

19

**Solution**   Suppose that the desired whole number $N$ has $n + 1$ digits, then $N = 4 \cdot 10^n + x$, where $x$ is an $n$-digit number. From assumptions in question

$$4(10x + 4) = 4 \cdot 10^n + x, \quad \text{i.e. } 39x = 4(10^n - 4) = 4 \cdot \underbrace{99\cdots9}_{n-1}6,$$

$$\therefore 13x = 4 \cdot \underbrace{33\cdots3}_{n-1}2, \quad \text{and } 13 \mid \underbrace{33\cdots3}_{n-1}2.$$

By checking the cases $n = 1, 2, \cdots$ one by one, it's easy to see that the minimal value of $n$ is 5:

$$33332 \div 13 = 2564. \quad \therefore x = 4 \times 2564 = 10256, \quad \text{and } N = 410256.$$

**Example 2.** (KIEV/1957) Find all two digit numbers such that each is divisible by the product of its two digits.

**Solution**   Let $\overline{xy} = 10x + y$ be a desired two digit number. Then there is a positive integer $k$ such that
$$10x + y = kxy,$$
so $y = (ky - 10)x$, i.e. $x \mid y$. Thus, $0 < x \leq y$.

If $x = y$, then $11x = kx^2$, so $kx = 11 = 11 \cdot 1$, i.e. $k = 11, x = 1 = y$ or $k = 1, x = 11 = y$ (N.A.). Thus, 11 is a solution.

If $x < y$, then $x \leq 4$ (otherwise, $y \geq 10$).

When $x = 4$, then $y = 8$. However, there is no positive integer $k$ such that $48 = 32k$, so $x \neq 4$.

When $x = 3$, $10x = (kx - 1)y$ gives $30 = (3k - 1)y$. Since $y \leq 9, 3 \mid y$ and $y \mid 30$, so $y = 6$. It is obvious that $36 = 2 \cdot 3 \cdot 6$, so 36 is the second solution.

When $x = 2$, then $20 = (2k - 1)y$. Since $y \leq 9, 2 \mid y$ and $y \mid 20$, so $y = 4$. $24 = 3 \cdot 2 \cdot 4$ verifies that 24 is the third solution.

When $x = 1$, then $10 = (k-1)y$. So $y = 2$ or 5. $12 = 6 \cdot 1 \cdot 2$ and $15 = 3 \cdot 1 \cdot 5$ indicate that 12 and 15 are solutions also.

Thus, the solutions are $11, 12, 15, 24, 36$.

**Example 3.** (CHNMO(P)/2002) A positive integer is called a "good number" if it is equal to four times of the sum of its digits. Find the sum of all good numbers.

**Solution**   If an one digit number $a$ is good number, then $a = 4a$, i.e. $a = 0$, so no one digit good number exists.

Let $\overline{ab} = 10a + b$ be a two digit good number, then $10a + b = 4(a + b)$ implies $2a = b$, so there are four good numbers $12, 24, 36, 48$, and their sum is 120.

Three digit good number $\overline{abc}$ satisfies the equation $100a + 10b + c = 4(a + b + c)$, i.e. $96a + 6b - 3c = 0$. Since $96a + 6b - 3c \geq 96 + 0 - 27 > 0$ always, so no solution for $(a, b, c)$, i.e. no three digit good number exists.

Since a number with $n$ ($n \geq 4$) digits must be not less than $10^{n-1}$, and the 4 times of the sum of its digits is not greater than $36n$. For $n \geq 4$,

$$10^{n-1} - 36n > 36(10^{n-3} - n) > 0,$$

so no $n$ digit good number exists if $n \geq 4$.

Thus, the sum of all good numbers is 120.

**Example 4.** (SSSMO(J)/2001) Let $\overline{abcdef}$ be a 6-digit integer such that $\overline{defabc}$ is 6 times the value of $\overline{abcdef}$. Find the value of $a + b + c + d + e + f$.

**Solution** From assumption in the question,

$$(1000)(\overline{def}) + \overline{abc} = 6[(1000)(\overline{abc}) + \overline{def}],$$
$$(994)(\overline{def}) = (5999)(\overline{abc}),$$
$$(142)(\overline{def}) = (857)(\overline{abc}).$$

Therefore $857 \mid (142)(\overline{def})$. Since 857 and 142 have no common factor greater than 1, so $857 \mid \overline{def}$. Since $2 \times 857 > 1000$ which is not a three digit number, so $\overline{def} = 857$. Thus, $\overline{abc} = 142$, and

$$a + b + c + d + e + f = 1 + 4 + 2 + 8 + 5 + 7 = 27.$$

**Example 5.** Prove that each number in the sequence 12, 1122, 111222, $\cdots$ is a product of two consecutive whole numbers.

**Solution** By using the decimal representation of a number with repeated digits, we have

$$\underbrace{11\cdots11}_{n}\underbrace{22\cdots22}_{n} = \frac{1}{9}(10^n - 1) \cdot \frac{2}{9}(10^n - 1) = \frac{1}{9}(10^n - 1)(10^n + 2)$$

$$= \left(\frac{10^n - 1}{3}\right) \cdot \left(\frac{10^n + 2}{3}\right) = \left(\frac{10^n - 1}{3}\right) \cdot \left(\frac{10^n - 1}{3} + 1\right) = A \cdot (A + 1),$$

where $A = \frac{1}{3}(10^n - 1) = \underbrace{33\cdots33}_{n}$ is a whole number. The conclusion is proven.

**Example 6.** (AIME/1986) In a parlor game, the magician asks one of the participants to think of a three digit number $\overline{abc}$ where $a, b$, and $c$ represent digits in base 10 in the order indicated. The magician then asks this person to form the numbers $\overline{acb}$, $\overline{bca}$, $\overline{bac}$, $\overline{cab}$, and $\overline{cba}$, to add these five numbers, and to reveal their sum, $N$. If told the value of $N$, the magician can identify the original number, $\overline{abc}$. Play the role of the magician and determine the $\overline{abc}$ if $N = 3194$.

**Solution**   Let $S = N + \overline{abc} = \overline{abc} + \overline{acb} + \overline{bca} + \overline{bac} + \overline{cab} + \overline{cba}$, then

$$
\begin{aligned}
S &= (100a + 10b + c) + (100a + 10c + b) + (100b + 10a + c) \\
&\quad + (100b + 10c + a) + (100c + 10a + b) + (100c + 10b + a) \\
&= 222(a + b + c).
\end{aligned}
$$

$3194 = N = 222(a + b + c) - \overline{abc}$ implies $222(a + b + c) = 3194 + \overline{abc} = 222 \times 14 + 86 + \overline{abc}$. Hence

(i)   $a + b + c > 14$;

(ii)   $86 + \overline{abc}$ is divisible by 222, i.e. $\overline{abc} + 86 = 222n$ for some positive integer $n$.

Since $222n \le 999 + 86 = 1085$, so $n \le \dfrac{1085}{222} < 5$, hence $n$ may be one of $1, 2, 3, 4$.

When $n = 1$, then $\overline{abc} = 222 - 86 = 136$, the condition (i) is not satisfied.

When $n = 2$, then $\overline{abc} = 444 - 86 = 358$, the conditions (i) and (ii) are satisfied.

When $n = 3$, then $\overline{abc} = 666 - 86 = 580$, the condition (i) is not satisfied.

When $n = 4$, then $\overline{abc} = 888 - 86 = 802$, the condition (i) is not satisfied.

Thus, $\overline{abc} = 358$.

**Example 7.** (MOSCOW/1940) Find all three-digit numbers such that each is equal to the sum of the factorials of its own digits.

**Solution**   Let $\overline{abc} = 100a + 10b + c$ be a desired three digit number.

$7! = 5040$ indicates that $a, b, c \le 6$, and further, if one of $a, b, c$ is 6, then

$$\overline{abc} > 6! = 720 \Rightarrow \text{one of } a, b, c \text{ is greater than } 6,$$

so $a, b, c \le 5$. Since $\overline{555} \ne 5! + 5! + 5!$, so $a, b, c$ cannot be all 5.

On the other hand, $4! + 4! + 4! = 72$ which is not a three digit number, so at least one of $a, b, c$ is 5. $\overline{abc} < 5! + 5! + 5! = 360$ implies that $a \le 3$.

When $a = 1$, then $145 = 1! + 4! + 5!$, so 145 is a desired number.

When $a = 2$, then $b, c$ must be 5. But $255 \ne 2! + 5! + 5!$, so no solution.

When $a = 3$, then $b, c$ must be 5. But $355 \ne 3! + 5! + 5!$, so no solution.

Thus, 145 is the unique solution.

**Example 8.** (IMO/1962) Find the smallest natural number $n$ which has the following properties:

(a) Its decimal representation has 6 as the last digit.

(b) If the last digit 6 is erased and placed in front of the remaining digits, the resulting number is four times as large as the original number $n$.

**Solution**   It is clear that $n$ is not a one-digit number. Let $n = 10x + 6$, where $x$ is a natural number of $m$ digits. Then

$$6 \cdot 10^m + x = 4(10x + 6) \Rightarrow 39x = 6 \cdot 10^m - 24 \Rightarrow 13x = 2 \cdot 10^m - 8,$$

so $13 \mid (2 \cdot 10^m - 8)$ for some $m$, i.e. the remainder of $2 \cdot 10^m$ is 8 when divided by 13. By long division, it is found that the minimum value of $m$ is 5. Thus,

$$x = \frac{2 \cdot 10^m - 8}{13} = \frac{199992}{13} = 15384, \quad n = 153846.$$

**Example 9.**  (KIEV/1963) Find all the three digit number $n$ satisfying the condition that if 3 is added, the sum of digits of the resultant number is $\dfrac{1}{3}$ of that of $n$.

**Solution**   Let $n = \overline{abc}$. By assumption the carry of digits must have happened when doing the addition $\overline{abc} + 3$, therefore $c \geq 7$.

By $S_0$ and $S_1$ we denote the sum of digits of $n$ and the resultant number respectively. There are three cases to be discussed below:

(i)  If $a = b = 9$, then $S_0 \geq 9 + 9 + 7 = 25$, but $S_1 = 1 + (c + 3 - 10) \leq 3$, a contradiction. Therefore the case is impossible.

(ii)  If $a < 9, b = 9$, then $S_0 = a + 9 + c$, $S_1 = a + 1 + (c + 3 - 10) = a + c - 6$. Therefore $3(a + c - 6) = a + 9 + c$, i.e. $2(a + c) = 27$, a contradiction. So no solution.

(iii)  If $b < 9$, then $S_0 = a + b + c$, $S_1 = a + (b + 1) + (c + 3 - 10) = a + b + c - 6$, it follows that $3(a + b + c - 6) = a + b + c$, i.e. $a + b + c = 9$, therefore $\overline{abc} = 108, \ 117, \ 207$.

Thus, $\overline{abc} = 108$ or $117$ or $207$.

## Testing Questions   (A)

1.  Prove that when $\overline{abc}$ is a multiple of 37, then so is the number $\overline{bca}$.

2.  (CMO/1970) Find all positive integers with initial digit 6 such that the integer formed by deleting this 6 is $1/25$ of the original integer.

3.  (SSSMO(J)/2000) Let $x$ be a 3-digit number such that the sum of the digits equals 21. If the digits of $x$ are reversed, the number thus formed exceeds $x$ by 495. What is $x$?

4.  Prove that each of the following numbers is a perfect square number:

$$729, \ 71289, \ 7112889, \ 711128889, \ \cdots.$$

5.  (ASUMO/1987) Find the least natural number $n$, such that its value will become $5n$ when its last digit is moved to the first place.

6.  (CHINA/2000) Given that a four digit number $n$ and all digits of $n$ have a sum 2001. Find $n$.

7.  (CHINA/1988) When $N = \underbrace{11\cdots11}_{1989\text{ digits}} \times \underbrace{11\cdots11}_{1989\text{ digits}}$ , what is the sum of all digits of $N$?

8.  (CHINA/1979) Given that a four digit number satisfies the following conditions: (i) when its units digit and hundreds digit are interchanged, and so does the tens digit and thousands digit, then the value of the number increases 5940. (ii) the remainder is 8 when it is divided by 9. Find the minimum odd number satisfying these conditions.

9.  (CHNMOL/1987) $x$ is a five digit odd number. When all its digits 5 are changed to 2 and all digits 2 are changed to 5, keeping all the other digits unchanged, a new five digit number $y$ is obtained. What is $x$ if $y = 2(x + 1)$?

10.  (MOSCOW/1954) Find the maximum value of the ratio of three digit number to the sum of its digits.

## Testing Questions    (B)

1.  (CHINA/1988) Find all the three digit numbers $n = \overline{abc}$ such that $n = (a + b + c)^3$.

2.  (ASUMO/1986) Given that the natural numbers $a, b, c$ are formed by the same $n$ digits $x$, $n$ digits $y$, and $2n$ digits $z$ respectively. For any $n \geq 2$ find the digits $x, y, z$ such that $a^2 + b = c$.

3.  (CHINA/1991) When a two digit number is divided by the number formed by exchanging the two digits, the quotient is equal to its remainder. Find the two digit number.

4.  (POLAND/1956) Find a four digit perfect square number, such that its first two digits and the last two digits are the same respectively.

5.  (RUSMO/1964) Find the maximum perfect square, such that after deleting its last two digits (which is assumed to be not all zeros), the remaining part is still a perfect square.

# Lecture 20

# Perfect Square Numbers

**Definition** A whole number $n$ is called a **perfect square number** (or shortly, **perfect square**), if there is an integer $m$ such that $n = m^2$.

## Basic Properties of Perfect Square Numbers

(I)     The units digit of a perfect square can be $0, 1, 4, 5, 6$ and $9$ only.
        It suffices to check the property for $0^2, 1^2, 2^2, \ldots, 9^2$.

(II)    If the prime factorization of a natural number $n$ is $p_1^{\alpha_1} p_2^{\alpha_2} \cdots p_k^{\alpha_k}$, then

$$n \text{ is a perfect square} \Leftrightarrow \text{each } \alpha_i \text{ is even} \Leftrightarrow \tau(n) \text{ is odd.}$$

where $\tau(n)$ denotes the number of positive divisors of $n$.

(III)   For any perfect square number $n$, the number of its tail zeros (i.e. the digit $0$s on its right end) must be even, since in the prime factorization of $n$ the number of factor $2$ and that of factor $5$ are both even.

(IV)    $n^2 \equiv 1$ or $0$ modulo $2, 3, 4$.
        It suffices to check the numbers of the forms $(2m)^2$ and $(2m + 1)^2$ by taking modulo $2$ and modulo $4$ respectively; the numbers of the forms of $(3m)^2$ and $(3m \pm 1)^2$ by taking modulo $3$.

(V)     $n^2 \equiv 0, 1$ or $4 \pmod 8$.
        It suffices to check the conclusion for $(4m \pm 1)^2$, $(4m)^2$, $(4m+2)^2$, where $m$ is any integer.

25

(VI)    An odd perfect square number must have an even tens digit (if one digit
        perfect squares $1^2$ and $3^2$ are considered as 01 and 09 respectively).
        It is easy to see the reasons: For $n > 3$, $n^2 = (10a + b)^2 = 100a^2 +$
        $20ab + b^2$. The number $100a^2 + 20ab$ has units digit 0 and an even tens
        digit. If $b$ is an odd digit, then the tens digit carried from $b^2$ must be even,
        so the tens digit of $n^2$ must be even.

(VII)   If the tens digit of $n^2$ is odd, then the units digit of $n^2$ must be 6.
        Continue the analysis in (VI). If the tens digit carried from $b^2$ is odd, then
        $b = 4$ or 6 only, so $b^2 = 16$ or 36, i.e. the units digit of $n^2$ must be 6.

(VIII)  There is no perfect square number between any two consecutive perfect
        square numbers $k^2$ and $(k + 1)^2$, where $k$ is any non-negative integer.
        Otherwise, there is a third integer between the two consecutive integers $k$
        and $k + 1$, however, it is impossible.

The basic problems involving perfect square numbers are (i) identifying if a
number is a perfect square; (ii) to find perfect square numbers under some conditions on perfect squares; (iii) to determine the existence of integer solution of
equations by use of the properties of perfect square numbers.

## Examples

**Example 1.** Prove that for any integer $k$, all the numbers of the forms $3k+2, 4k+$
$2, 4k + 3, 5k + 2, 5k + 3, 8n \pm 2, 8n \pm 3, 8n + 7$ cannot be perfect squares.

**Solution**   $3k + 2 \equiv 2 \pmod 3$, $4k + 2 \equiv 2 \pmod 4$, $4k + 3 \equiv 3 \pmod 4$
implies they cannot be perfect squares.

$5k + 2$ has units digit 2 or 7 and $5k + 3$ has units digit 3 or 8, so they cannot
be perfect squares also.

All the numbers $8n \pm 2, 8n \pm 3, 8n + 7$ have no remainders 0, 1 or 4 modulo
8, so they cannot be perfect squares.

**Example 2.** Prove that any positive integer $n \geq 10$ cannot be a perfect square
number if it is formed by the same digits.

**Solution**   If the used digit is odd, then the conclusion is proven by the property (VI).

If the used digit is 2 or 8, the conclusion is obtained at once from the property
(I).

If the used digit is 6, the conclusion is obtained by the property (VII).

Finally, if the used digit is 4, let $n^2 = \underbrace{44\cdots44}_{k}$ with $k \geq 2$, then $n = 2m$ i.e.

$n^2 = 4m^2$, so $m^2 = \underbrace{11\cdots 11}_{k}$, a contradiction from above discussion.

Thus, the conclusion is proven for all the possible cases.

**Example 3.** (AHSME/1979) The square of an integer is called a perfect square number. If $x$ is a perfect square number, then its next one is

(A) $x+1$, (B) $x^2+1$, (C) $x^2+2x+1$, (D) $x^2+x$, (E) $x+2\sqrt{x}+1$.

**Solution** Since $x \geq 0$, so $x = (\sqrt{x})^2$, and its next perfect square is $(\sqrt{x}+1)^2 = x + 2\sqrt{x} + 1$, the answer is (E).

**Example 4.** Prove that the sum of 1 and the product of any four consecutive integers must be a perfect square, but the sum of any five consecutive perfect squares must not be a perfect square.

**Solution** Let $a, a+1, a+2, a+3$ are four consecutive integers. Then

$$a(a+1)(a+2)(a+3) + 1 = [a(a+3)][(a+1)(a+2)] + 1$$
$$= (a^2+3a)(a^2+3a+2) + 1 = [(a^2+3a+1) - 1][(a^2+3a+1) + 1] + 1$$
$$= (a^2+3a+1)^2,$$

so the first conclusion is proven.

Now let $(n-2)^2, (n-1)^2, n^2, (n+1)^2, (n+2)^2$ be any five consecutive perfect squares. Then

$$(n-2)^2 + (n-1)^2 + n^2 + (n+1)^2 + (n+2)^2 = 5n^2 + 10 = 5(n^2+2).$$

If $5(n^2+2)$ is a perfect square, then $5 \mid (n^2+2)$, so $n^2$ has 3 or 7 as its units digit, but this is impossible. Thus, the second conclusion is also proven.

**Example 5.** (CHNMOL/1984) In the following listed numbers, the one which must not be a perfect square is

(A) $3n^2 - 3n + 3$, (B) $4n^2 + 4n + 4$, (C) $5n^2 - 5n - 5$,
(D) $7n^2 - 7n + 7$, (E) $11n^2 + 11n - 11$.

**Solution** $3n^2 - 3n + 3 = 3(n^2 - n + 1)$ which is $3^2$ when $n = 2$;
$5n^2 - 5n - 5 = 5(n^2 - n - 1) = 5^2$ when $n = 3$;
$7n^2 - 7n + 7 = 7(n^2 - n + 1) = 7^2$ when $n = 3$;
$11n^2 + 11n - 11 = 11(n^2 + n - 1) = 11^2$ when $n = 3$.

Therefore (A), (C), (D) and (E) are all not the answer. On the other hand,

$$(2n+1)^2 = 4n^2 + 4n + 1 < 4n^2 + 4n + 4 < 4n^2 + 8n + 4 = (2n+2)^2$$

implies that $4n^2 + 4n + 4$ is not a perfect square. Thus, the answer is (B).

**Example 6.** (CHINA/2002) Given that five digit number $\overline{2x9y1}$ is a perfect square number. Find the value of $3x + 7y$.

**Solution**　We use estimation method to determine $x$ and $y$. Let $A^2 = \overline{2x9y1}$.
Since $141^2 = 19881 < A^2$ and $175^2 = 30625 > A^2$, so $141^2 < A^2 < 175^2$.
the units digit of $A^2$ is 1 implies that the units digit of $A$ is 1 or 9 only. Therefore
it is sufficient to check $151^2, 161^2, 171^2, 159^2, 169^2$ only, so we find that

$$161^2 = 25921$$

satisfies all the requirements, and other numbers cannot satisfy all the requirements. Thus,

$$x = 5, y = 2, \quad \text{so that } 3x + 7y = 15 + 14 = 29.$$

**Example 7.** (CHNMOL/2004) Find the number of the pairs $(x, y)$ of two positive
integers, such that $N = 23x + 92y$ is a perfect square number less than or equal
to 2392.

**Solution**　$N = 23x + 92y = 23(x + 4y)$ and 23 is a prime number implies
that $x + 4y = 23m^2$ for some positive integer $m$, so

$$N = 23^2 m^2 \le 2392 \implies m^2 \le \frac{2392}{23^2} = \frac{104}{23} < 5.$$

Hence $m^2 = 1$ or 4, i.e. $m = 1$ or 2.

When $m^2 = 1$, then $x + 4y = 23$ or $x = 23 - 4y$. Since $x, y$ are two positive
integers, so $y = 1, 2, 3, 4, 5$ and $x = 19, 15, 11, 7, 3$ correspondingly.

When $m^2 = 4$, then $x + 4y = 92$ or $x = 92 - 4y$, so $y$ can take each positive
integer value from 1 through 22, and $x$ then can take the corresponding positive
integer values given by $x = 92 - 4y$.

Thus, the number of qualified pairs $(x, y)$ is $5 + 22$, i.e. 27.

**Example 8.** (CHINA/2006) Prove that 2006 cannot be expressed as the sum of
ten odd perfect square numbers.

**Solution**　We prove by contradiction. Suppose that 2006 can be expressed as
the sum of ten odd perfect square numbers, i.e.

$$2006 = x_1^2 + x_2^2 + \cdots + x_{10}^2,$$

where $x_1, x_2, \ldots, x_{10}$ are all odd numbers. When taking modulo 8 to both sides,
the left hand side is 6, but the right hand side is 2, a contradiction! Thus, the
assumption is wrong, and the conclusion is proven.

**Example 9.** (CHINA/1991) Find all the natural number $n$ such that $n^2 - 19n + 91$
is a perfect square.

**Solution** (i) When $n > 10$, then $n - 9 > 0$, so

$$n^2 - 19n + 91 = n^2 - 20n + 100 + (n - 9) = (n - 10)^2 + (n - 9) > (n - 10)^2,$$

and

$$n^2 - 19n + 91 = n^2 - 18n + 81 + (10 - n) < (n - 9)^2,$$

so $(n - 10)^2 < n^2 - 19n + 91 < (n - 9)^2$, which implies that $n^2 - 19n + 91$ is not a perfect square.

(ii) When $n < 9$, then

$$n^2 - 19n + 91 = (10 - n)^2 + (n - 9) < (10 - n)^2$$

and

$$n^2 - 19n + 91 = (9 - n)^2 + 10 - n > (9 - n)^2,$$

so $(9 - n)^2 < n^2 - 19n + 91 < (10 - n)^2$, i.e. $n^2 - 19n + 91$ cannot be a perfect square.

(iii) When $n = 9$, then $n^2 - 19n + 91 = (10 - 9)^2 = 1$; when $n = 10$, then $n^2 - 19n + 91 = (10 - 9)^2 = 1$.

Thus, $n^2 - 19n + 91$ is a perfect square if and only if $n = 9$ or $10$.

## Testing Questions (A)

1. Determine if there is a natural number $k$ such that the sum of the two numbers $3k^2 + 3k - 4$ and $7k^2 - 3k + 1$ is a perfect square.

2. If $(x - 1)(x + 3)(x - 4)(x - 8) + m$ is a perfect square, then $m$ is

    (A) 32,      (B) 24,      (C) 98,      (D) 196.

3. (CHINA/2006) If $n + 20$ and $n - 21$ are both perfect squares, where $n$ is a natural number, find $n$.

4. If the sum of 2009 consecutive positive integers is a perfect square, find the minimum value of the maximal number of the 2009 numbers.

5. (ASUMO/1972) Find the maximal integer $x$ such that $4^{27} + 4^{1000} + 4^x$ is a perfect square.

6. (KIEV/1970) Prove that for any positive integer $n$, $n^4 + 2n^3 + 2n^2 + 2n + 1$ is not a perfect square.

7.  (ASUMO/1973) If a nine digit number is formed by the nine non-zero digits, and its units digit is 5, prove that it must not be a perfect square.

8.  (KIEV/1975) Prove that there is no three digit number $\overline{abc}$, such that $\overline{abc} + \overline{bca} + \overline{cab}$ is a perfect square.

9.  (Canada/1969) Prove that the equation $a^2 + b^2 - 8c = 6$ has no integer solution.

10. (BMO/1991) Show that if $x$ and $y$ are positive integers such that $x^2 + y^2 - x$ is divisible by $2xy$, then $x$ is a perfect square.


# Testing Questions   (B)

1.  Find the number of ordered pairs $(m, n)$ of two integers with $1 \le m, n \le 99$, such that $(m + n)^2 + 3m + n$ is a perfect square number.

2.  Given that $p$ is a prime number, and the sum of all positive divisors of $p^4$ is a perfect square. Find the number of such primes $p$.

3.  (CHINA/1992) If $x$ and $y$ are positive integers, prove that the values of $x^2 + y + 1$ and $y^2 + 4x + 3$ cannot both be perfect squares at the same time.

4.  (IMO/1986) Let $d$ be any positive integer not equal to 2, 5, or 13. Show that one can find distinct $a$, $b$ in the set $\{2, 5, 13, d\}$ such that $ab - 1$ is not a perfect square.

5.  (BMO/1991) Prove that the number $3^n + 2 \times 17^n$, where $n$ is a non-negative integer, is never a perfect square.

# Lecture 21

# Pigeonhole Principle

## Basic Forms of Pigeonhole Principle

Principle I:      When $m + 1$ pigeons enter $m$ pigeonholes ($m$ is a positive integer), there must be at least one hole having more than 1 pigeon.

Principle II:      When $m + 1$ pigeons enter $n$ pigeonholes, there must be one hole having at least $\left\lfloor \dfrac{m}{n} \right\rfloor + 1$ pigeons.

Principle III:      When infinitely many elements are partitioned into finitely many sets, there must be at least one set containing infinitely many elements.

It is easy to understand the Pigeonhole Principle, and the above three forms can be proven at once by contradiction. But it does not mean that the use of the principle is easy. Pigeonhole principle has various applications. By use of the principle to prove the existence of some case, essentially is to classify all the possible cases into a few or only several classes, then use proof by contradiction to show the desired conclusion. So the central problem for applying the principle is doing classification of the possible cases, i.e. to find out different but appropriate "pigeonholes" for different problems. However, up to the present, there is still no unified method to find out the "pigeonholes". The examples below will demonstrate several methods for getting pigeonholes.

## Examples

**Example 1.** (CHINA/2003) In a bag, there are some balls of the same size that are colored by 7 colors, and for each color the number of balls is 77. At least how many balls are needed to be picked out at random to ensure that one can obtain 7 groups of 7 balls each such that in each group the balls are homochromatic?

31

**Solution**    For this problem, it is natural to let each color be one pigeonhole, and a ball drawn be a pigeon. At the first step, for getting a group of 7 balls with the same color, at least 43 balls are needed to be picked out from the bag at random, since if only 42 balls are picked out, there may be exactly 6 for each color. By pigeonhole principle, there must be one color such that at least $\lfloor 42/7 \rfloor + 1 = 7$ drawn balls have this color.

Next, after getting the first group, it is sufficient to pick out from the bag another 7 balls for getting 43 balls once again. Then, by the same reason, the second group of 7 homochromatic drawn balls can be obtained.

Repeating this process for 6 times, the 7 groups of 7 homochromatic balls are obtained. Thus, the least number of drawn balls is $43 + 6 \times 7 = 85$.

**Example 2.** (SSSMO(J)/2001) A bag contains 200 marbles. There are 60 red ones, 60 blue ones, 60 green ones and the remaining 20 consist of yellow and white ones. If marbles are chosen from the bag without looking, what is the smallest number of marbles one must pick in order to ensure that, among the chosen marbles, at least 20 are of the same colour?

**Solution**    When 77 marbles are chosen, there may be 19 red, 19 blue, 19 green and 20 yellow and white.

If 78 marbles are chosen at random, the number of yellow and white ones among them is at most 20. Therefore there are at least 58 marbles of red, blue or green colors. According to the Pigeonhole Principle, the number of drawn marbles of some color is not less than $\left\lfloor \dfrac{57}{3} \right\rfloor + 1 = 20$, i.e. at least 20. Thus, the smallest number of marbles to be picked is 78.

In this problem, a color is taken as a pigeonhole, and then a drawn marble is taken as a pigeon.

**Example 3.** (CHINA/2001) If 51 numbers are arbitrarily taken out from the first 100 natural numbers $\{1, 2, \ldots, 100\}$, prove that there must be two numbers drawn such that one is a multiple of the other.

**Solution**    Let $A = \{1, 2, 3, \ldots, 100\}$. There are a total of 50 odd numbers in the set $A$: $2k - 1$, $k = 1, 2, 3, \ldots, 50$. Any number in $A$ can be written uniquely in the form $2^m q$, where $m$ is a non-negative integer, and $q$ is an odd number.

When all the numbers in $A$ are partitioned according to $q$ into 50 classes, then each number in $A$ must belong to a unique class. Now for any 51 numbers in $A$, by the Pigeonhole Principle, there must be two of them coming from a same class, and these two numbers must satisfy the requirement: one is multiple of another since they have the same $q$.

In this problem, a pigeonhole is the set of all numbers in $A$ with same maximum odd factor $q$, since any two numbers in the pigeonhole have required property.

**Example 4.** (SSSMO/1999) Determine the maximum number of elements of a subset $L$ of $\{1, 2, 3, ..., 1999\}$ such that the difference of any two distinct elements of $L$ is not equal to 4.

**Solution** Let $A = \{1, 2, 3, \cdots, 1999\}$. At first, by modulo 4, we can partition $A$ into four subsets $L_0$, $L_1$, $L_2$ and $L_3$, where

$$L_i = \{n \in A : n \equiv i \pmod{4}\}, \qquad \text{for } i = 0, 1, 2, 3.$$

Then there are 500 numbers in each of $L_1$, $L_2$, $L_3$ but 499 numbers in $L_0$.

For the set $L_1$, arrange all the numbers in an ascending order, and consider the 250 pairs formed by its numbers $(1, 5)$, $(9, 13)$, $(17, 21)$, $\ldots$, $(1993, 1997)$.

By Pigeonhole Principle, if more than 250 numbers in $L_1$ are selected out as part of $L$, then there must be some two numbers coming from one of the above pairs, so their difference is 4. Thus, any more than 250 numbers in $L_1$ cannot be put in $L$. However, all numbers on odd numbered places or all numbers on even places can be chosen. Thus at most 250 numbers can be chosen from $L_1$ as a subset of $L$, and the same analysis works for each of $L_2$, $L_2$ and $L_0$ (in $L_0$ all the 250 numbers which are odd multiples of 4 can be chosen to put in $L$). Therefore the maximum number of the elements in $L$ is 1000.

For applying the pigeonhole principle, the listed pairs take the role of pigeonholes, and the chosen numbers are the pigeons.

The congruence relation is a method commonly used for classifying integers. This method is also often used for applying pigeonhole principle, as shown in the following examples.

**Example 5.** (SSSMO/1990) Given any $2n - 1$ positive integers, prove that there are $n$ of them whose sum is divisible by $n$ for (1) $n = 3$; (2) $n = 9$.

**Solution** (1) When $n = 3$ then $2n - 1 = 5$. Partition the five numbers according to the remainders modulo 3 into three classes: $C_0, C_1$, and $C_2$.

(i) If one of the three classes contains no number, i.e. five numbers are in two classes, by the pigeonhole principle, there must be one class containing at least 3 numbers, then any three numbers coming from a same class must have a sum divisible by 3;

(ii) If each class contains at least one number, then no class contains three or more numbers. But from each class take a number, the three numbers must have a sum divisible by 3.

(2) For $n = 9$, then $2n - 1 = 17$ numbers are given. The result of (1) implies that, from each five of them three numbers can be selected out such that their sum is divisible by 3, so 5 groups $(n_1, n_2, n_3), (n_4, n_5, n_6), \cdots, (n_{13}, n_{14}, n_{15})$ can be obtained sequentially from the 17 numbers, such that their sums $s_1, s_2, \cdots s_5$

are all divisible by 3. Let $s_i = 3m_i, i = 1, 2, \ldots, 5$ (where $m_i$ are all positive integers). Then, still from the result of (1), three numbers, say $m_1, m_2, m_3$ can be selected from the five $m_i$ such that $m_1 + m_2 + m_3 = 3k$ for some positive integer $k$. Thus,

$$n_1 + n_2 + \cdots + n_9 = s_1 + s_2 + s_3 = 3(m_1 + m_2 + m_3) = 9k,$$

which is divisible by 9.

**Example 6.** Prove that for any given 50 positive integers, it is always possible to select out four numbers $a_1, a_2, a_3$ and $a_4$ from them, such that $(a_2 - a_1)(a_4 - a_3)$ is a multiple of 2009.

**Solution** First of all, note that $2009 = 49 \times 41$. Consider the 50 remainders of the 50 given integers modulo 49, by the pigeonhole principle, theere must be two numbers selected from the 50 integers, denoted by $a_1$ and $a_2$, such that $a_1$ and $a_2$ are congruent modulo 49, so $a_2 - a_1$ is divisible by 49.

Next, by same reason, it must be possible that two numbers $a_3$ and $a_4$ can be selected from the remaining 48 numbers such that $a_4 - a_3$ is divisible by 41.

Thus, $49 \cdot 41 \mid (a_2 - a_1)(a_4 - a_3)$, i.e. $2009 \mid (a_2 - a_1)(a_4 - a_3)$.

**Example 7.** Prove that in a set containing $n$ positive integers there must be a subset such that the sum of all numbers in it is divisible by $n$.

**Solution** Let the $n$ positive integers be $a_1, a_2, \ldots, a_n$. Consider $n$ new positive integers:

$$b_1 = a_1, \quad b_2 = a_1 + a_2, \quad \cdots, b_n = a_1 + a_2 + \cdots + a_n.$$

Then all the $n$ values are distinct. When some of $b_1, b_2, \ldots, b_n$ is divisible by $n$, the conclusion is proven. Otherwise, if all $b_i$ are not divisible by $n$, then their remainders are all not zero, i.e. at most they can take $n - 1$ different values. By the pigeonhole principle, there must be $b_i$ and $b_j$ with $i < j$ such that $b_j - b_i \neq 0$ is divisible by $n$. Since $b_j - b_i = a_{i+1} + a_{i+2} + \cdots + a_j$ is a sum of some given numbers, the conclusion is proven.

**Example 8.** (CHINA/1993) If five points are taken at random from within a square of side 1, prove that there must be two of them such that the distance between them is not greater than $\dfrac{\sqrt{2}}{2}$.

**Solution** By connecting the two midpoints of two opposite sides, the square is partitioned into four smaller congruent squares of side $\dfrac{1}{2}$. By the pigeonhole principle, there must be one such square containing at least 2 points.

Since the small square is covered by the circle with same center and a radius of $\sqrt{2}/4$, so the distance between any two points in the circle is not greater than the diameter, i.e. $\sqrt{2}/2$. The conclusion is proven.

**Example 9.** (HUNGARY/1947) There are six points in the space such that any three are not collinear. If any two of them are connected by a segment, and each segment is colored by one of red color or blue color, prove that there must be at least **one** triangles formed by three points and segments joining them, such that the three sides are of the same color.

**Solution** Let the six points be $A_0$, $A_1$, $A_2$, $A_3$, $A_4$ and $A_5$. Consider the segments joining these points. We use a real segment to denote a red segment and a dot segment to denote a blue segment, as shown in the diagram.

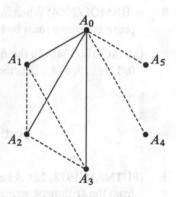

Consider the five segments starting from $A_0$: $A_0A_1, A_0A_2, A_0A_3, A_0A_4, A_0A_5$. Each of them is colored by red color or blue color. By Pigeon-hole Principle, among them there must be at least three with the same color. Without loss of generality, it can be assumed that three are red, and they are $A_0A_1, A_0A_2, A_0A_3$.

Now consider three sides of the triangle $A_1A_2A_3$. When one side is red, then there is a red triangle, i.e. its three sides are all red. Otherwise, the three sides of $\triangle A_1A_2A_3$ are all blue,, then the conclusion is also true.

## Testing Questions    (A)

1. Prove that among any $n + 1$ integers, there must be at least two which are congruent modulo $n$.

2. Let $1 \le a_1 < a_2 < a_3 < \cdots < a_{n+1} \le 2n$ be $n + 1$ integers, where $n \ge 1$. Prove that there must be two of them, $a_i < a_j$, such that $a_i \mid a_j$.

3. Prove that in the 2009 numbers 1, 11, $\cdots$, $\underbrace{111\cdots111}_{2009}$, there is one that is divisible by 2009.

4. 19 points are taken randomly inside an equilateral triangle of area 1 m². Prove that among the triangles formed by these points, there must be one with area not greater than $\dfrac{1}{9}$ m².

5.  (CHINA/2002) Prove that among any seven distinct integers, there must be two such that their sum or difference is divisible by 10.

6.  (CHINA/2003) In a chess board of dimension $4 \times 28$, every small square is colored by one of red color, blue color or yellow color. Prove that under any coloring there must be a rectangular region with four angles of same color.

7.  (CHINA/1996) Given that each of the nine straight lines cuts a square to two trapezia with a ratio of $2/3$. Prove that at least three of the nine lines pass through a same point.

8.  (CHNMOL/2004) When inserting $n+1$ points on the segment $OA$ at random, prove that there must be two of them with a distance not greater than $1/n$.

9.  (MOSCOW/1946) In the first 100000001 terms of the Fibonacci's sequence $0, 1, 1, 2, 3, 5, 8, \ldots$, is there a term ending with at least four zeros?

# Testing Questions    (B)

1.  (PUTNAM/1978) Let $A$ be a set formed by choosing 20 numbers arbitrarily from the arithmetic sequence $1, 4, 7, \ldots, 100$. Prove that there must be two numbers in $A$ such that their sum is 104.

2.  (CHINA/2005) On the blackboard some student has written 17 natural numbers, and their units digits are inside the set $\{0, 1, 2, 3, 4\}$. Prove that one can always select out 5 numbers from them such that their sum is divisible by 5.

3.  (IMO/1972) Prove that from a set of ten distinct two-digit numbers (in the decimal system), it is possible to select two disjoint subsets whose members have the same sum.

4.  (IMO/1964) Seventeen people correspond by mail with one another — each one with all the rest. In their letters only three different topics are discussed. Each pair of correspondents deals with only one of these topics. Prove that there are at least three people who write to each other about the same topic.

5.  (IMO/1978) An international society has its members from six different countries. The list of members contains 1978 names, numbered $1, 2, \ldots, 1978$. Prove that there is at least one member whose number is the sum of the numbers of two members from his own country, or twice as large as the number of one member from his own country.

# Lecture 22

# $\lfloor x \rfloor$ and $\{x\}$

**Definition 1**  For any real number $x$, the largest integer less than or equal to $x$, denoted by $\lfloor x \rfloor$, is called the **integer part** of $x$. When $x$ is considered as a real variable, the function $f(x) = \lfloor x \rfloor$, $x \in \mathbb{R}$ is called the **Gaussian function**.

**Definition 2**  For any real number $x$, the value $x - \lfloor x \rfloor$, denoted by $\{x\}$, is called the **decimal part** (or **fractional part**) of $x$.

**Some Basic Properties of $\lfloor x \rfloor$ and $\{x\}$**

- $0 \le \{x\} < 1$, and $\{x\} = 0$ if and only if $x$ is an integer.

- $x - 1 < \lfloor x \rfloor \le x < \lfloor x \rfloor + 1$.

- For any $n \in \mathbb{Z}$, $\lfloor n + x \rfloor = n + \lfloor x \rfloor$.

- $\lfloor -x \rfloor = \begin{cases} -\lfloor x \rfloor - 1 & \text{if } x \text{ is not an integer,} \\ -\lfloor x \rfloor & \text{if } x \text{ is an integer.} \end{cases}$

- $\lfloor x + y \rfloor \ge \lfloor x \rfloor + \lfloor y \rfloor$ for any $x, y \in \mathbb{R}$. In general, for $x_1, \ldots, x_n \in \mathbb{R}$,
$$\lfloor x_1 + x_2 + \cdots + x_n \rfloor \ge \lfloor x_1 \rfloor + \lfloor x_2 \rfloor + \cdots + \lfloor x_n \rfloor.$$

- $\lfloor xy \rfloor \ge \lfloor x \rfloor \cdot \lfloor y \rfloor$, where $x, y \ge 0$. In general, for $x_1, x_2, \cdots, x_n \ge 0$,
$$\lfloor x_1 x_2 \cdots x_n \rfloor \ge \lfloor x_1 \rfloor \cdot \lfloor x_2 \rfloor \cdot \cdots \cdot \lfloor x_n \rfloor.$$

- $\left\lfloor \dfrac{x}{n} \right\rfloor = \left\lfloor \dfrac{\lfloor x \rfloor}{n} \right\rfloor$ for $n \in \mathbb{N}$, $x \in \mathbb{R}$.

37

**Theorem I.** *(Hermite Identity) For any $x \in \mathbb{R}$ and $n \in \mathbb{N}$,*

$$\lfloor x \rfloor + \left\lfloor x + \frac{1}{n} \right\rfloor + \left\lfloor x + \frac{2}{n} \right\rfloor + \cdots + \left\lfloor x + \frac{n-1}{n} \right\rfloor = \lfloor nx \rfloor.$$

**Proof.**    Define the auxiliary function

$$f(x) = \lfloor x \rfloor + \left\lfloor x + \frac{1}{n} \right\rfloor + \left\lfloor x + \frac{2}{n} \right\rfloor + \cdots + \left\lfloor x + \frac{n-1}{n} \right\rfloor - \lfloor nx \rfloor.$$

Then it suffices to show $f(x) = 0$ identically. Since

$$
\begin{aligned}
f\left(x + \frac{1}{n}\right) &= \left\lfloor x + \frac{1}{n} \right\rfloor + \left\lfloor x + \frac{2}{n} \right\rfloor + \cdots + \lfloor x + 1 \rfloor - \lfloor nx + 1 \rfloor \\
&= \lfloor x \rfloor + \left\lfloor x + \frac{1}{n} \right\rfloor + \left\lfloor x + \frac{2}{n} \right\rfloor + \cdots + \left\lfloor x + \frac{n-1}{n} \right\rfloor - \lfloor nx \rfloor \\
&= f(x).
\end{aligned}
$$

so $f(x)$ is a periodic function with a period $\dfrac{1}{n}$, hence it is enough to show $f(x) = 0$ for $0 \le x < \dfrac{1}{n}$, and this is obvious from the definition of $f$.    $\square$

**Theorem II.** *(Legendre's Theorem) In the prime factorization of the product $n! = 1 \times 2 \times 3 \times \cdots \times n$, the index of a prime factor $p$ is given by*

$$\left\lfloor \frac{n}{p} \right\rfloor + \left\lfloor \frac{n}{p^2} \right\rfloor + \left\lfloor \frac{n}{p^3} \right\rfloor + \cdots.$$

**Proof.**    In $n!$ the index of its prime factor $p$ is the sum of the indices of prime factor $p$ in the numbers $1, 2, \ldots, n$. Since in the $n$ numbers 1 through $n$ there are $\left\lfloor \dfrac{n}{p} \right\rfloor$ numbers containing at least one factor $p$, $\left\lfloor \dfrac{n}{p^2} \right\rfloor$ numbers containing at least one factor $p^2$, ..., so above sum can count the total number of factor $p$ in $n!$, the conclusion is proven.    $\square$

Besides the problems about $\lfloor x \rfloor$ and $\{x\}$ themselves, the concepts of $\lfloor x \rfloor$ and $\{x\}$ established the connection between $x$ and them, so the basic problems involving $\lfloor x \rfloor$ and $\{x\}$ contain also those from $x$ get $\lfloor x \rfloor$ and $\{x\}$, or conversely, from $\lfloor x \rfloor$ and $\{x\}$ get $x$, i.e. solving equations with $\lfloor x \rfloor$ and $\{x\}$ for $x$.

Another kind of related problems is to give some theoretical discussions involving $\lfloor x \rfloor$ and $\{x\}$, as indicated in Theorem I and Theorem II. But here we will give some examples only belonging to the first two kinds of problems.

**Examples**

**Example 1.** Solve the equation $2\lfloor x \rfloor = x + 2\{x\}$.

**Solution** To reduce the number of variables, $x = \lfloor x \rfloor + \{x\}$ yields the equation
$$2\lfloor x \rfloor = \lfloor x \rfloor + 3\{x\},$$
$$\therefore \lfloor x \rfloor = 3\{x\} < 3.$$

When $\lfloor x \rfloor = 0, 1, 2$ respectively, then $\{x\} = 0, \dfrac{1}{3}, \dfrac{2}{3}$ correspondingly, so the solutions are $x = 0, \dfrac{4}{3}, \dfrac{8}{3}$.

**Example 2.** (CMO/1999) Find all real solutions to the equation $4x^2 - 40\lfloor x \rfloor + 51 = 0$.

**Solution** To reduce the number of variables in the given equation, $x \geq \lfloor x \rfloor$ gives an inequality in $x$:

$$0 = 4x^2 - 40\lfloor x \rfloor + 51 \geq 4x^2 - 40x + 51 = (2x - 3)(2x - 17),$$

so $\dfrac{3}{2} \leq x \leq \dfrac{17}{2}$. It implies that $\lfloor x \rfloor$ may be $1, 2, \ldots, 8$.

When $\lfloor x \rfloor = 1$, the equation becomes $4x^2 + 11 = 0$, no real solution for $x$.

When $\lfloor x \rfloor = 2$, the equation becomes $4x^2 - 29 = 0$, $x = \dfrac{\sqrt{29}}{2}$, which has the integer part 2, hence it is a solution.

When $\lfloor x \rfloor = 3$, the equation becomes $4x^2 - 69 = 0$, $x = \dfrac{\sqrt{69}}{2}$ which has no integer part 3, a contradiction, hence no solution for $x$.

When $\lfloor x \rfloor = 4$, the equation becomes $4x^2 - 109 = 0$, $x = \dfrac{\sqrt{109}}{2} > 5$, a contradiction, hence no solution.

When $\lfloor x \rfloor = 5$, the equation becomes $4x^2 - 149 = 0$, $x = \dfrac{\sqrt{149}}{2} > 6$, hence no solution.

When $\lfloor x \rfloor = 6$, the equation becomes $4x^2 - 189 = 0$, $x = \dfrac{\sqrt{189}}{2}$, which has integer part 6, hence it is a solution.

When $\lfloor x \rfloor = 7$, the equation becomes $4x^2 - 229 = 0$, $x = \dfrac{\sqrt{229}}{2}$, which has integer part 7, hence it is a solution.

When $\lfloor x \rfloor = 8$, the equation becomes $4x^2 - 269 = 0$, $x = \dfrac{\sqrt{269}}{2}$, which has integer part 8, hence it is a solution.

Thus, the solutions are $\dfrac{\sqrt{29}}{2}, \dfrac{\sqrt{189}}{2}, \dfrac{\sqrt{229}}{2}, \dfrac{\sqrt{269}}{2}$.

**Example 3.** (CHINA/1986) Find the maximum positive integer $k$, such that

$$\frac{1001 \cdot 1002 \cdots \cdots 1985 \cdot 1986}{11^k}$$

is an integer.

**Solution**   Let $N = \dfrac{1001 \cdot 1002 \cdots \cdots 1985 \cdot 1986}{11^k}$, then

$$N = \frac{1000! \cdot 1001 \cdot 1002 \cdots \cdots 1985 \cdot 1986}{11^k (1000!)} = \frac{1986!}{11^k (1000!)}.$$

The highest power of 11 in 1986! is

$$\left\lfloor \frac{1986}{11} \right\rfloor + \left\lfloor \frac{1986}{11^2} \right\rfloor + \left\lfloor \frac{1986}{11^3} \right\rfloor = 180 + 16 + 1 = 197.$$

The highest power of 11 in 1000! is $\left\lfloor \dfrac{1000}{11} \right\rfloor + \left\lfloor \dfrac{1000}{11^2} \right\rfloor = 90 + 8 = 98$, so the maximum value of $k$ is given by

$$k = 197 - 98 = 99.$$

**Example 4.** (SSSMO/2002) Determine the number of real solutions of

$$\left\lfloor \frac{x}{2} \right\rfloor + \left\lfloor \frac{2x}{3} \right\rfloor = x.$$

**Solution**   The given equation indicates that any solution $x$ must be an integer. Let $x = 6q + r$, where $r = 0, 1, 2, 3, 4, 5$ and $q$ is an integer. Then the given equation becomes

$$q + \left\lfloor \frac{r}{2} \right\rfloor + \left\lfloor \frac{2r}{3} \right\rfloor = r.$$

(i)      $r = 0$ gives $q = 0$, so $x = 0$ is a solution.
(ii)     $r = 1$ gives $q = 1$, so $x = 7$ is a solution.
(iii)    $r = 2$ gives $q = 0$, so $x = 2$ is a solution.
(iv)     $r = 3$ gives $q = 0$, so $x = 3$ is a solution.
(v)      $r = 4$ gives $q = 0$, so $x = 4$ is a solution.
(vi)     $r = 5$ gives $q = 0$, so $x = 5$ is a solution.
   Thus, there are a total of 6 real solutions.

**Example 5.** (SSSMO(J)/2001) Let $x, y, z$ be three positive real numbers such that

$$x + \lfloor y \rfloor + \{z\} = 13.2 \tag{22.1}$$
$$\lfloor x \rfloor + \{y\} + z = 14.3 \tag{22.2}$$
$$\{x\} + y + \lfloor z \rfloor = 15.1 \tag{22.3}$$

where $\lfloor a \rfloor$ denotes the greatest integer $\leq a$ and $\{b\}$ denotes the fractional part of $b$ (for example, $\lfloor 5.4 \rfloor = 5$, $\{4.3\} = 0.3$). Find the value of $x$.

**Solution** $(22.1) + (22.2) + (22.3)$ yields $2(x + y + z) = 42.6$, i.e.

$$x + y + z = 21.3 \tag{22.4}$$

$(22.4) - (22.1)$ gives $\{y\} + \lfloor z \rfloor = 8.1$, therefore $\lfloor z \rfloor = 8$ and $\{y\} = 0.1$.

Then $(22.2)$ gives $\lfloor x \rfloor + z = 14.2$, so $\{z\} = 0.2$ and $z = 8.2$ which gives $\lfloor x \rfloor = 6$.

$(22.1)$ gives $x + \lfloor y \rfloor = 13$, so $x$ is an integer, i.e. $x = \lfloor x \rfloor = 6$.

**Example 6.** (CHINA/1988) Let $S = \lfloor \sqrt{1} \rfloor + \lfloor \sqrt{2} \rfloor + \cdots + \lfloor \sqrt{1988} \rfloor$. Find $\lfloor \sqrt{S} \rfloor$.

**Solution** For any positive integer $k$ and $x$, the following relations are equivalent

$$\lfloor \sqrt{x} \rfloor = k \Leftrightarrow k^2 \leq x < (k + 1)^2 \Leftrightarrow x \in [k^2, k^2 + 2k],$$

so $2k + 1$ values of $x$ satisfy the relation. Since $44^2 = 1936 < 1988 < 2025 = 45^2$ and $1988 - 1936 + 1 = 53$,

$$
\begin{aligned}
S &= 1(3) + 2(5) + 3(7) + \cdots + 43(87) + 44(53) \\
&= 2(1^2 + 2^2 + \cdots + 43^2) + (1 + 2 + \cdots + 43) + 44(53) \\
&= \frac{43 \cdot 44 \cdot 87}{3} + \frac{43 \cdot 44}{2} + 2332 = 54868 + 946 + 2332 = 58146.
\end{aligned}
$$

Since $240^2 = 57600 < 241^2 = 58081 < S < 242^2 = 58564$, $\lfloor \sqrt{S} \rfloor = 241$.

**Example 7.** (CHINA/1986) Evaluate the sum $S = \sum_{k=1}^{502} \left\lfloor \dfrac{305k}{503} \right\rfloor$.

**Solution** If each of two positive real numbers $x, y$ is not an integer but $x + y$ is an integer, then $\lfloor x \rfloor + \lfloor y \rfloor = x + y - 1$ since $\{x\} + \{y\} = 1$. Since $\dfrac{305k}{503} + \dfrac{305(503 - k)}{503} = 305$ for $1 \leq k \leq 502$,

$$
\begin{aligned}
S &= \left\lfloor \frac{305}{503} \right\rfloor + \left\lfloor \frac{305 \cdot 2}{503} \right\rfloor + \cdots + \left\lfloor \frac{305 \cdot 502}{503} \right\rfloor \\
&= \left( \left\lfloor \frac{305 \cdot 1}{503} \right\rfloor + \left\lfloor \frac{305 \cdot 502}{503} \right\rfloor \right) + \cdots + \left( \left\lfloor \frac{305 \cdot 251}{503} \right\rfloor + \left\lfloor \frac{305 \cdot 252}{503} \right\rfloor \right) \\
&= 304 \cdot 251 = 76304.
\end{aligned}
$$

**Example 8.** (PUTNAM/1986) What is the units digit of $\left\lfloor \dfrac{10^{20000}}{10^{100} + 3} \right\rfloor$?

**Solution**   Let $n = 10^{100}$. Then

$$\frac{10^{20000}}{10^{100} + 3} = \frac{n^{200}}{n + 3} = \frac{(n^2)^{100} - (3^2)^{100}}{n + 3} + \frac{9^{100}}{n + 3}$$

$$= \frac{(n^2 - 3^2)M}{n + 3} + \frac{9^{100}}{n + 3} = (n - 3)M + \frac{9^{100}}{n + 3}.$$

Since $9^{100} < n$, we have

$$\left\lfloor \frac{n^{200}}{n + 3} \right\rfloor = (n - 3)M = \frac{n^{200} - 9^{100}}{n + 3} = \frac{10^{20000} - 81^{50}}{10^{100} + 3}.$$

Since the units digit of $10^{20000} - 81^{50}$ is 9 and the units digit of $10^{100} + 3$ is 3, the units digit of the quotient must be 3.

**Example 9.** (MOSCOW/1951) Given some positive integers less than 1951, where each two have a lowest common multiple greater than 1951. Prove that the sum of the reciprocals of these numbers is less than 2.

**Solution**   Let the given positive integers be $a_1, a_2, \ldots, a_n$. Since any two of them have an L.C.M. greater than 1951, so each of $1, 2, 3, \ldots, 1951$ cannot be divisible by any two of $a_1, a_2, \ldots, a_n$.

Therefore the number $M$ of numbers in the set $\{1, 2, \ldots, 1951\}$ which is divisible by one of $a_1, a_2, \ldots, a_n$ is given by

$$M = \left\lfloor \frac{1951}{a_1} \right\rfloor + \left\lfloor \frac{1951}{a_2} \right\rfloor + \cdots + \left\lfloor \frac{1951}{a_n} \right\rfloor,$$

then $M \le 1951$.

On the other hand, the inequalities $\left\lfloor \dfrac{1951}{a_i} \right\rfloor > \dfrac{1951}{a_i} - 1$ for $i = 1, 2, \ldots, n$ hold, so

$$\left( \frac{1951}{a_1} - 1 \right) + \left( \frac{1951}{a_2} - 1 \right) + \cdots + \left( \frac{1951}{a_1} - 1 \right) < 1951,$$

$$\frac{1951}{a_1} + \frac{1951}{a_2} + \cdots + \frac{1951}{a_n} < 1951 + n < 2 \cdot 1951,$$

$$\therefore \frac{1}{a_1} + \frac{1}{a_2} + \cdots + \frac{1}{a_n} < 2.$$

**Example 10.** (CHINA/1992) Given that real number $a > 1$, the natural number $n \ge 2$, and the equation $\lfloor ax \rfloor = x$ has exactly $n$ distinct real solution. Fine the range of $a$.

**Solution** From assumptions each solution $x$ must be an integer and $x \geq 0$. Since

$$ax = \lfloor a \rfloor x + \{a\}x,$$

so the given equation becomes

$$x = \lfloor ax \rfloor = \lfloor a \rfloor x + \lfloor \{a\}x \rfloor. \tag{22.5}$$

Since $\lfloor a \rfloor \geq 1$, (22.5) holds if and only if

$$\lfloor a \rfloor = 1 \quad \text{and} \quad \{a\}x < 1. \tag{22.6}$$

Since $x = 0$ must be a solution, so the equation $\lfloor ax \rfloor = x$ has exactly $n - 1$ positive solution. Since $\{a\}x < 1$ implies $\{a\}x' < 1$ if $0 < x' < x$, so the positive integer solutions of $\{a\}x < 1$ must be $x = 1, 2, \ldots, n-1$, hence $\{a\} < 1/(n-1)$. On the other hand, since any integer $\geq n$ is not a solution of (22.5), so $\{a\} \geq 1/n,$. Thus, $1/n \leq \{a\} < 1/(n - 1)$. Since $\lfloor a \rfloor = 1$, so

$$1 + \frac{1}{n} \leq a < 1 + \frac{1}{n - 1}.$$

## Testing Questions    (A)

1.  (PUTNAM/1948) If $n$ is a positive integer, prove that $\lfloor \sqrt{n} + \sqrt{n + 1} \rfloor = \lfloor \sqrt{4n + 2} \rfloor$.

2.  (KIEV/1972) Solve equation $\lfloor x^3 \rfloor + \lfloor x^2 \rfloor + \lfloor x \rfloor = \{x\} - 1$.

3.  (CMO/1975) Solve equation $\lfloor x \rfloor^2 = \{x\} \cdot x$.

4.  (ASUMO/1987) Find all solutions to the equation $x^2 - 8\lfloor x \rfloor + 7 = 0$.

5.  (SWE/1982) Given that $n$ is a natural number, how many roots of the equation

$$x^2 - \lfloor x^2 \rfloor = (x - \lfloor x \rfloor)^2$$

are in the interval $1 \leq x \leq n$?

6.  (CHNMOL/1987) Solve equation $\lfloor 3x + 1 \rfloor = 2x - \frac{1}{2}$, and find the sum of all roots.

7.  (ASUMO/1989) Find the minimum natural number $n$, such that the equation $\left\lfloor \dfrac{10^n}{x} \right\rfloor = 1989$ has integer solution $x$.

8.  (AIME/1985) How many of the first 1000 positive integers can be expressed in the form
    $$\lfloor 2x \rfloor + \lfloor 4x \rfloor + \lfloor 6x \rfloor + \lfloor 8x \rfloor,$$
    where $x$ is a real number, and $\lfloor z \rfloor$ denotes the greatest integer less than or equal to $z$?

9.  (ASUMO/1980) How many different non-negative integers are there in the sequence
    $$\left\lfloor \frac{1^2}{1980} \right\rfloor, \left\lfloor \frac{2^2}{1980} \right\rfloor, \left\lfloor \frac{3^2}{1980} \right\rfloor, \cdots, \left\lfloor \frac{1980^2}{1980} \right\rfloor?$$

10  (SSSMOJ/2000/Q30) Find the total number of integers $n$ between 1 and 10000 (both inclusive) such that $n$ is divisible by $\lfloor \sqrt{n} \rfloor$. Here $\lfloor \sqrt{n} \rfloor$ denotes the largest integer less than or equal to $\sqrt{n}$.

## Testing Questions   (B)

1.  (MOSCOW/1981) For $x > 1$ be the inequality $\left\lfloor \sqrt{\lfloor \sqrt{x} \rfloor} \right\rfloor = \left\lfloor \sqrt{\sqrt{x}} \right\rfloor$ must true?

2.  (CMO/1987) For every positive integer $n$, show that
    $$\lfloor \sqrt{n} + \sqrt{n+1} \rfloor = \lfloor \sqrt{4n+1} \rfloor = \lfloor \sqrt{4n+2} \rfloor = \lfloor \sqrt{4n+3} \rfloor.$$

3.  (USSR/1991) Solve the equation $\lfloor x \rfloor \{x\} + x = 2\{x\} + 10$.

4.  (CHNMOL/1993) Find the last two digits of the number $\left\lfloor \dfrac{10^{93}}{10^{31} + 3} \right\rfloor$ (Write down the tens digit first, then write down the units digit).

5.  (ASUMO/1992) Solve the equation $x + \dfrac{92}{x} = \lfloor x \rfloor + \dfrac{92}{\lfloor x \rfloor}$.

# Lecture 23

# Diophantine Equations (I)

## Definitions

For a polynomial with integral coefficients of multi-variables $f(x_1, \ldots, x_n)$, the problem for finding integer solutions $(x_1, \ldots, x_n)$ of the equation

$$f(x_1, \ldots, x_n) = 0$$

is called the **Diophantine problem**, and the equation $f(x_1, \ldots, x_n) = 0$ is called the **Diophantine equation**.

As usual, in a Diophantine equation or a system of Diophantine equations, the number of unknown variables is more than the number of equations, so the solutions may not be unique /finite, or say, the number of solutions is more than one as usual. Hence, based on the uncertainty of solutions, this kind of equations are also called **indefinite equations** in countries. Besides, the **Diophantine problem** is also called the problem for finding **integer solutions of equations**.

In this chapter the discussion is confined to linear equations and the systems of linear equations. The simplest and typical equation is

$$ax + by = c, \tag{23.1}$$

where $a, b, c$ are constant integers.

**Theorem I.** *The equation (23.1) has no integer solution if* $\gcd(a, b) \nmid c$.

**Proof.** Otherwise, if $(x, y)$ is an integer solution of (23.1), then left hand side is divisible by $\gcd(a, b)$, whereas the right hand side is not, a contradiction. $\square$

If $(a, b) \mid c$, then, after dividing by $(a, b)$, the both sides of (23.1) becomes

$$a'x + b'y = c', \tag{23.2}$$

45

where $a'$, $b'$, $c'$ are integers and $(a', b') = 1$. Therefore, below it is always assumed that the Eq.(23.1) satisfies the condition $(a, b) = 1$.

**Theorem II.** *When $(a, b) = 1$, the equation (23.1) always has at least one integer solution.*

**Proof.** Consider the case $c = 1$ first. For the equation $ax + by = 1$, the conclusion is clear if one of $a, b$ is 0 or 1. therefore we may assume that $b > 1$.

When each of the $b$ numbers $a, 2a, 3a, \ldots, b \cdot a$ is divided by $b$, then the remainders are distinct. Otherwise, if $ia \equiv ja \pmod{b}$ for some $1 \leq i < j \leq b$, then $b \mid (j - i)a$, so $b \mid (j - i)$. However $1 \leq j - i < b$, a contradiction.

Thus, there must be $x \in \{1, 2, \ldots, b\}$ such that $xa \equiv 1 \pmod{b}$, i.e. $xa - 1 = -yb$ for some integer $y$. Thus $ax + by = 1$.

When $c \neq 1$ and $(x_0, y_0)$ is an integer solution for $ax + by = 1$, then $(cx_0, cy_0)$ is an integer solution of (23.1). The Theorem II is proven. $\square$

**Theorem III.** *If $x_0$, $y_0$ is a special integer solution of the Equation (23.1), then the general solution of (23.1) is given by*

$$\begin{cases} x = x_0 + bt \\ y = y_0 - at, \end{cases} \quad \forall t \in \mathbb{Z}. \tag{23.3}$$

**Proof.** Let $(x, y)$ be any solution of (23.1) differ from the given solution $(x_0, y_0)$. Then

$$ax + by = c, \tag{23.4}$$
$$ax_0 + by_0 = c. \tag{23.5}$$

(23.4) − (23.5) gives $a(x - x_0) = -b(y - y_0)$, so $\dfrac{x - x_0}{b} = \dfrac{y - y_0}{-a}$. Since $b \mid a(x - x_0)$ and $a \mid b(y - y_0)$ implies that $b \mid (x - x_0)$ and $a \mid (y - y_0)$, so each side of the last equality is an integer. Letting it be $t$, it is follows that

$$x - x_0 = bt \quad \text{and} \quad y - y_0 = -at,$$

so the equalities in (23.3) are proven. $\square$

For solving a Diophantine equation based on the Theorem III, we can find the general solution if a special solution is given or can be found easily, then determine the range or values of $t$ according to the given conditions in question.

For solving a system of Diophantine equations or an equation with more than two variables, substitution method can be used for simplifying the equation.

**Examples**

**Example 1.** (CHINA/2001) Find all positive integer solutions to the equation $12x + 5y = 125$.

**Solution**  $12x = 5(25 - y)$ indicates $5 \mid x$. Let $x = 5$, then $5y = 65$ gives $y = 13$, so $(5, 13)$ is a special solution. By the formula for general solution, it is obtained that

$$x = 5 + 5t \quad \text{and} \quad y = 13 - 12t, \quad \text{where } t \text{ is an integer.}$$

Since $x \geq 1$, so $t \geq 0$. But $y \geq 1$ implies $t \leq 1$, so $t = 0$ or $1$.

When $t = 0$, the solution is $x = 5, y = 13$. When $t = 1$, then $x = 10, y = 1$. Thus, the equation has exactly two solutions.

**Example 2.** Find the general solution of the Diophantine equation $17x + 83y = 5$.

**Solution**  It is not obvious to find a special solution. Here substitution can be used to simplify the equation. Since

$$x = \frac{5 - 83y}{17} = -4y + \frac{5 - 15y}{17},$$

let $k = \dfrac{5 - 15y}{17}$, then $y = \dfrac{5 - 17k}{15} = -k + \dfrac{5 - 2k}{15}$. Let $m = \dfrac{5 - 2k}{15}$, then $k = \dfrac{5 - 15m}{2} = 2 - 7m + \dfrac{1 - m}{2}$. Let $t = \dfrac{1 - m}{2}$, then $m = 1 - 2t$. By substituting back these relations, it follows that

$$k = 2 - 7(1 - 2t) + t = -5 + 15t,$$
$$y = -(-5 + 15t) + (1 - 2t) = 6 - 17t,$$
$$\therefore x = -4(6 - 17t) + (-5 + 15t) = -29 + 83t.$$

**Example 3.** Given that the positive integers $x > 1$ and $y$ satisfies the equation $2007x - 21y = 1923$. Find the minimum value of $2x + 3y$.

**Solution**  Simplify the given equation to $669x = 7y + 641$. Changing its form to

$$669(x - 1) = 7(y - 4),$$

since $x - 1 > 0$, so $y > 4$ and the minimum positive values of $x$ and $y$ are given by $x - 1 = 7, y - 4 = 669$. Thus,

$$2x_{\min} + 3y_{\min} = 2 \cdot 8 + 3 \cdot 673 = 2035.$$

**Example 4.** Find all the integer solutions of the equation $25x + 13y + 7z = 6$.

**Solution**   Let $25x + 13y = U$, then $U + 7z = 6$. Consider $U$ as a constant at the moment to solve the equation

$$25x + 13y = U.$$

Since $(-U, 2U)$ is a special solution for $(x, y)$, the general solution is obtained:

$$x = -U + 13t_1, \qquad y = 2U - 25t_1, \quad t_1 \in \mathbb{Z}.$$

Next, solve the equation $U + 7z = 6$ for $(U, z)$. Since $(-1, 1)$ is a special solution, the general solution is given by

$$U = -1 + 7t_2, \quad \text{and} \quad z = 1 - t_2, \quad t_2 \in \mathbb{Z}.$$

By substituting the expression of $U$ into those for $x$ and $y$, the general solution is obtained:

$$x = 1 + 13t_1 - 7t_2, \quad y = -2 - 25t_1 + 14t_2, \quad z = 1 - t_2, \quad t_1, t_2 \in \mathbb{Z}.$$

**Example 5.** (SSSMO(J)/2002) The digits $a, b$ and $c$ of a three-digit number $\overline{abc}$ satisfy $49a + 7b + c = 286$. Find the three-digit number $\overline{abc}$.

**Solution**   By taking modulo 7 to both sides of the given equation, it follows that

$$c \equiv 6 \pmod 7.$$

Since $c$ is a digit, so $c = 6$. Then the given equation becomes $7a + b = 40$ or $7a = 40 - b$. Since $31 \le 40 - b \le 40$, and there is only one number 35 divisible by 7 in this interval, so $b = 5, a = 5$, i.e. $\overline{abc} = 556$.

**Example 6.** (CHINA/2007) Given that the equation $\dfrac{4}{3}x - a = \dfrac{2}{5}x + 140$ has a positive integer solution, where $a$ is a parameter. Find the minimum positive integer value of $a$.

**Solution**   The given equation produces $a = 14\left(\dfrac{x}{15} - 10\right)$. When $a$ is a positive integer, then $15 \mid x$ and $x > 150$, therefore $x_{\min} = 165$ and $a_{\min} = 14$.

**Example 7.** (AHSME/1989) Given that $n$ is a positive integer, and the equation $2x + 2y + z = n$ has a total of 28 positive integer solutions for $(x, y, z)$. Then the value of $n$ is
    (A) 14 or 15;   (B) 15 or 16;   (C) 16 or 17;   (D) 17 or 18;   (E) 18 or 19.

**Solution**   Let $u = x + y$, then $2u + z = n$. Consider $u \ge 2$ as a constant at the moment for solving $x$ and $y$. For the equation $x + y = u$, $(u - 1, 1)$ is a special solution, so $x = u - 1 + t_1, y = 1 - t_1, t_1 \in \mathbb{Z}$ is the general solution.

Since $x \geq 1, y \geq 1$, so $2 - u \leq t_1 \leq 0$, i.e. $t_1$ has $u - 1$ permitted values for each given value $u \geq 2$.

Now consider the equation $2u + z = n$.

When $n$ is even, then $z$ is even also, the equation becomes $u + \dfrac{z}{2} = \dfrac{n}{2}$. The value of $u$ can take from 2 to $\frac{n}{2} - 1$. Thus, the number of positive solutions is

$$(2 - 1) + (3 - 1) + \cdots + (\frac{n}{2} - 2) = 28.$$

Since $1 + 2 + \cdots + 7 = 28$, so $\dfrac{n}{2} - 2 = 7$, i.e. $n = 18$.

When $n = 2k + 1$, then $u$ can take the value from 2 to $\frac{n-1}{2}$, so the number of positive solutions is

$$1 + 2 + \cdots + (\frac{n - 1}{2} - 1) = 28.$$

From $\dfrac{n - 1}{2} - 1 = 7$ we have $n = 16 + 1 = 17$. Thus, the answer is (D).

**Example 8.** Find the integer solutions of the equation $13x - 7y = 0$ satisfying the condition $80 < x + y < 120$.

**Solution**　Since $13x = 7y$ has integer solution $(0, 0)$, so the general solution is

$$x = 7t, \quad y = 13t, \quad t \in \mathbb{Z}.$$

Then $x + y = 20t$, so $80 < x + y < 120 \Leftrightarrow 4 < t < 6$, i.e. $t = 5$. Thus,

$$x = 35, y = 65$$

is the unique desired solution.

**Example 9.** (ASUMO/1988) There are two piles of pebbles, pile (A) and pile (B). When 100 pebbles are moved from (A) to (B), then the number of pebbles in (B) is double of that in (A). However, if some are moved from (B) to (A), then the number of pebbles in (A) is five times more than that in (B). What is the minimum possible number of pebbles in (A), and find the number of pebbles in (B) in that case.

**Solution**　Let $x$ and $y$ be the numbers of pebbles in the piles (A) and (B) respectively. When $z$ pebbles are moved from (B) to (A), then the given conditions in question give

$$2(x - 100) = y + 100, \tag{23.6}$$
$$x + z = 6(y - z). \tag{23.7}$$

(23.6) gives $y = 2x - 300$, so from (23.7) it follows that

$$11x - 7z = 1800$$

or

$$4x + 7(x - z) = 1800. \tag{23.8}$$

Both sides taking modulo 4 yields $4 \mid (x - z)$, so $x - z = 4t, t \in \mathbb{Z}$. Then (23.8) implies $4x + 28t = 1800$ or $x + 7t = 450$. Thus, the general solution for $(x, y, z)$ is

$$x = 450-7t, \quad y = 2(450-7t)-300 = 600-14t, \quad z = (450-7t)-4t = 450-11t.$$

From $x, y, z \geq 0$, it is obtained that $t \leq \dfrac{450}{11} < 41$, so $t \leq 40$. When $t$ takes its maximum possible value then $x$ is its minimum, so

$$x_{min} = 450 - 280 = 170.$$

In that case, $y = 600 - 560 = 40, z = 450 - 440 = 10$, so there are 40 pebbles in (B).

**Example 10.** Find all triples $(x, y, z)$ of three non-negative integers satisfying the system of equations

$$5x + 7y + 5z = 37 \tag{23.9}$$
$$6x - y - 10z = 3. \tag{23,10}$$

**Solution**   By eliminating a variable from the system, the question will become one with two variables. $2 \times (23.9) + (23.10)$ yields

$$16x + 13y = 77.$$

Since $16(x - 4) + 13(y - 1) = 0$, then $x = 4, y = 1$ is a special solution. So the general solution is

$$x = 4 + 13t, \quad y = 1 - 16t, \quad t \in \mathbb{Z}.$$

Since $y \geq 0$, so $t \leq 0$. But $x \leq 0$ implies $t \geq 0$, so $t = 0$ is the unique permitted value of $t$. Thus, $x = 4, y = 1$, and $z = 2$ from (23.9).

## Testing Questions (A)

1. (AHSME/1992) If $k$ is a positive integer such that the equation in $x$

$$kx - 12 = 3k$$

   has an integer root, then the number of such $k$ is

   (A) 3;   (B) 4;   (C) 5;   (D) 6;   (E) 7.

2. (CHINA/1990) An integer solution of the equation $1990x - 1989y = 1991$ is

   (A) $x = 12785, y = 12768$;   (B) $x = 12785, y = 12770$;

   (C) $x = 11936, y = 11941$;   (D) $x = 13827, y = 12623$.

3. (SSSMO(J)/2002) Two positive integers $A$ and $B$ satisfy $\dfrac{A}{11} + \dfrac{B}{3} = \dfrac{17}{33}$. Find the value of $A^2 + B^2$.

4. (CHINA/1997) A four digit number has remainder 13 when divided by 16, and has remainder 122 when divided by 125. Find the minimum value of such four digit numbers.

5. A dragonfly has six feet and a spider has 8 feet. Given that a certain group of dragonflies and spiders have in total 46 feet, find the number of dragonflies and the number of spiders.

6. Given that $x$ 1-cent coins, $y$ 2-cent coins, and $z$ 5-cent coins have a total value of 10 dollars. Find the number of solutions $(x, y, z)$.

7. If a four digit number and the sum of its all digits have a sum 2006, find the four digit number.

8. (CHINA/1997) $m, n$ are integers satisfying $3m + 2 = 5n + 3$ and $30 < 3m + 2 < 40$, find the value of $mn$.

9. (Ancient Question) In an ancient chicken market, each rooster is sold for 5 coins, each hen for 3 coins and each chick for $1/3$ coin. Someone has 100 coins to buy 100 chickens, how many roosters, hens and chicks can a man purchase out of a total cost of 100 coins?

10. (SSSMO(J)/1997) Suppose $x, y$ and $z$ are positive integers such that $x > y > z > 663$ and $x, y$ and $z$ satisfy the following:

$$\begin{aligned} x + y + z &= 1998 \\ 2x + 3y + 4z &= 5992. \end{aligned}$$

   Find the values of $x, y$ and $z$.

## Testing Questions　(B)

1. (RUSMO/1983) Given that a pile of 100 small weights have a total weight of 500 g, and the weight of a small weight is 1g, 10 g or 50 g. Find the number of each kind of weights in the pile.

2. (ASUMO/1988) Prove that there are infinitely many positive integer solutions $(x, y, z)$ to the equation $x - y + z = 1$, such that $x, y, z$ are distinct, and any two of them have a product which is divisible by the remaining number.

3. $a, b$ are two relatively prime positive integers. Prove that the equation

$$ax + by = ab - a - b$$

has no non-negative integer solution.

4. Prove that for relatively prime two positive integers $a$ and $b$, the equation

$$ax + by = c$$

must have non-negative integer solution if $c > ab - a - b$.

5. (KIEV/1980) Multiply some natural number by 2 and then plus 1, and then carry out this operation on the resultant number, and so on. After repeating 100 times of such operations, whether the resulting number is divisible by (i) 1980?　(ii) by 1981?

# Roots and Discriminant of Quadratic Equation
$$ax^2 + bx + c = 0$$

**Definition 1**  The equality $ax^2 + bx + c = 0$ is called a **quadratic equation**, where $a, b, c$ are real constant coefficients with $a \neq 0$, $x$ is the unknown variable.

**Definition 2**  A real number $\alpha$ is called a **root** or **solution** of the quadratic equation $ax^2 + bx + c = 0$ if it satisfies the equation, i.e. $a(\alpha)^2 + b\alpha + c = 0$.

**Definition 3**  For the quadratic equation, the value $\Delta = b^2 - 4ac$ is called the **discriminant** of the equation.

**Basic Methods for Finding Roots of $ax^2 + bx + c = 0$**

(I)   Based on $ax^2 + bc + c = a\left(x + \dfrac{b}{2a}\right)^2 + \dfrac{4ac - b^2}{4a}$, the roots $\alpha$ and $\beta$ can be given by

$$\alpha = \frac{-b - \sqrt{\Delta}}{2a}, \quad \beta = \frac{-b + \sqrt{\Delta}}{2a}.$$

(II)   By factorizing $ax^2 + bx + c$ to the form $a(x - a_1)(x - a_2)$, the roots then are $a_1$ and $a_2$. All the methods for factorizations can be used here, including multiplication formulae, factor theorem and observation method, etc.

(III)   For quadratic equations with absolute values, it is needed to convert them to normal equations by substitution or by partitioning the range of $x$ piecewise to remove the absolute signs.

**Relation between Discriminant and Existence of Real Roots**

(I)      Use $\Delta$ to determine the existence of real roots without solving the equation:
        (i)    $\Delta > 0 \Leftrightarrow$ the equation has two distinct real roots.
        (ii)   $\Delta = 0 \Leftrightarrow$ the equation has two equal real roots.
        (iii)  $\Delta < 0 \Leftrightarrow$ the equation has no real roots.

(II)   Geometrical explanation of the relation is as follows: since the real roots of the quadratic equation are the $x$-coordinates of the points of the image $\Gamma$ of the function $y = ax^2 + bx + c$ (which is a parabola) on the $x$-axis,
        (i)    $\Delta > 0 \Leftrightarrow$ the curve $\Gamma$ intersects the $x$-axis at two distinct points.
        (ii)   $\Delta = 0 \Leftrightarrow$ the curve $\Gamma$ is tangent to the $x$-axis at one point.
        (iii)  $\Delta < 0 \Leftrightarrow$ the curve $\Gamma$ and the $x$-axis have no point of intersection.

**Examples**

**Example 1.** (CHINA/2006) Solve the equation $2006x^2 + 2007x + 1 = 0$

    **Solution**  Let $f(x) = 2006x^2 + 2007x + 1 = 0$. By observation, $f(-1) = 2006 - 2007 + 1 = 0$, so $f(x)$ has the factor $(x + 1)$, and it is easy to find the second factor (by the synthetic division),

$$2006x^2 + 2007x + 1 = (x + 1)(2006x + 1),$$

so the two real roots are $-1$ and $-\dfrac{1}{2006}$.

**Example 2.** Solve the equations in $x$:
(i)      $(a^2 - 1)x + a(x^2 - 1) = a^2(x^2 - x + 1)$.
(ii)    $x^2 - 2(a^2 + b^2)x + (a^2 - b^2)^2 = 0$.

    **Solution**  The given equations have parameters, so the discussion on parameters is necessary.
(i)      The given equation can be written in the form

$$a(a - 1)x^2 - (2a^2 - 1)x + a(a + 1) = 0. \qquad (24.1)$$

Note that the equation is not quadratic if $a = 0$ or $1$. If $a = 0$, then $x = 0$ is the unique solution. If $a = 1$, then $x = 2$ is the unique solution. If $a(a - 1) \neq 0$, then left hand side of (24.1) can be factorized as

$$[ax - (a + 1)][(a - 1)x - a] = 0,$$

i.e. its two real roots are $x_1 = \dfrac{a + 1}{a}, x_2 = \dfrac{a}{a - 1}$.

(ii)     By completing squares, the given equation can be written in the form $[x - (a^2 + b^2)]^2 = 4a^2b^2$, so

$$x = a^2 + b^2 \pm 2ab = (a \pm b)^2.$$

**Note:**   This example indicates that it is not always convenient to use the formula for roots to solve an quadratic equation, in particular, if the equation contains parameters.

**Example 3.** Given that the equation $(x - 19)(x - 97) = p$ has real roots $r_1$ and $r_2$. Find the minimum real root of the equation $(x - r_1)(x - r_2) = -p$.

**Solution**   The problem can be solved by manipulating the left hand side of the equation.

$(x - 19)(x - 97) = p$ has real roots $r_1$ and $r_2$
$\Leftrightarrow (x - 19)(x - 97) - p = 0$ has real root $r_1$ and $r_2$
$\Rightarrow (x - 19)(x - 97) - p = (x - r_1)(x - r_2)$ for any $x$
$\Rightarrow (x - r_1)(x - r_2) + p = (x - 19)(x - 97)$ for any $x$
$\Rightarrow$ the roots of $(x - r_1)(x - r_2) + p = 0$ are 19 and 97
$\Rightarrow$ the roots of $(x - r_1)(x - r_2) = -p$ are 19 and 97.

Thus, the minimum root of $(x - r_1)(x - r_2) = -p$ is 19.

**Example 4.** (CHINA/2003) Let $a$ be the minimum root of the equation $x^2 - 3|x| - 2 = 0$, find the value of $-\dfrac{1}{a}$.

**Solution I**   It is clear that 0 is not a root. $(-a)^2 - 3|-a| - 2 = a^2 - 3|a| - 2 = 0$ implies that $-a$ is also a root, and $a < -a$ yields $a < 0$.

Thus, it suffices to find the maximum positive roots, so we solve the equation $x^2 - 3x - 2 = 0$. By the formula for roots, the root $\dfrac{3 + \sqrt{17}}{2} > 0$ is obtained. Thus,

$$a = -\frac{3 + \sqrt{17}}{2},$$

hence

$$-\frac{1}{a} = \frac{2}{3 + \sqrt{17}} = \frac{2(\sqrt{17} - 3)}{17 - 9} = \frac{\sqrt{17} - 3}{4}.$$

**Solution II**   Let $y = |x|$, then the given equation becomes $y^2 - 3y - 2 = 0$. By using the formula for roots,

$$|x| = y_1 = \frac{3 + \sqrt{17}}{2} \quad \text{(the negative root is not acceptable)}.$$

Thus, $a = -\dfrac{3 + \sqrt{17}}{2}$, and the rest is the same as Solution I.

**Example 5.** Solve the equation $|x^2 - 3x - 4| = |x - 2| - 1$.

　　**Solution**　$x^2 - 3x - 4 = 0$ has two roots $x = -1$ and $x = 4$, and $x - 2 = 0$ gives $x = 2$. So the number axis should be partitioned into four parts by the partition points $-1, 2$, and $4$.

(i)　　When $x \le -1$, the equation becomes $x^2 - 3x - 4 = 1 - x$, so $x^2 - 2x - 5 = 0$, the root is $x = 1 - \sqrt{6}$.

(ii)　　When $-1 < x \le 2$, the equation becomes $-(x^2 - 3x - 4) = 1 - x$, so $x^2 - 4x + 5 = 0$, no solution since $\varDelta < 0$.

(iii)　　When $2 < x \le 4$, the equation becomes $-(x^2 - 3x - 4) = x - 3$, so $x^2 - 2x - 7 = 0$, $x = 1 + 2\sqrt{2}$.

(iv)　　When $4 < x$, the equation becomes $x^2 - 3x - 4 = x - 3$, so $x^2 - 4x - 1 = 0$, $x = 2 + \sqrt{5}$.

　　Thus, the solutions are $x_1 = 1 - \sqrt{6}$, $x_2 = 1 + 2\sqrt{2}$, $x_3 = 2 + \sqrt{5}$.

**Example 6.** Given that $a$ is a root of the equation $x^2 - x - 3 = 0$. Evaluate $\dfrac{a^3 + 1}{a^5 - a^4 - a^3 + a^2}$.

　　**Solution**　$a^2 - a - 3 = 0$ yields $a^2 - a = 3$. On the other hand,

$$
\begin{aligned}
a^3 + 1 &= (a + 1)(a^2 - a + 1), \\
a^5 - a^4 - a^3 + a^2 &= a^2(a^3 - a^2 - a + 1) = a^2[(a^2(a - 1) - (a - 1)] \\
&= a^2(a - 1)(a^2 - 1) = a^2(a + 1)(a - 1)^2 \\
&= (a + 1)(a^2 - a)^2,
\end{aligned}
$$

it follows that

$$
\frac{a^3 + 1}{a^5 - a^4 - a^3 + a^2} = \frac{a^2 - a + 1}{(a^2 - a)^2} = \frac{4}{3^2} = \frac{4}{9}.
$$

　　For quadratic equations with parameters, the information on roots can conversely determine the range of the parameters, as shown in the following examples.

**Example 7.** Given that the equation $x^2 - (2a + b)x + (2a^2 + b^2 - b + \frac{1}{2}) = 0$ has two real roots. Find the values of $a$ and $b$.

　　**Solution**　$\varDelta \ge 0$ implies that

$$
\begin{aligned}
&(2a + b)^2 - 4(2a^2 + b^2 - b + \tfrac{1}{2}) \ge 0, \\
&4a^2 + 3b^2 - 4ab - 4b + 2 \le 0, \\
&(2a - b)^2 + 2(b - 1)^2 \le 0, \quad \therefore a = 2, \ b = 1.
\end{aligned}
$$

**Example 8.** (CHINA/2005) Given that the equations $x^2 - ax + 3 - b = 0$ has two distinct real roots, $x^2 + (6 - a)x + 6 - b = 0$ has two equal real roots, and $x^2 + (4 - a)x + 5 - b = 0$ has no real roots. Then the ranges of $a$ and $b$ are
    (A) $2 < a < 4, 2 < b < 5$,     (B) $1 < a < 4, 2 < b < 5$,
    (C) $2 < a < 4, 1 < b < 5$,     (D) $1 < a < 4, 1 < b < 5$.

**Solution** The assumptions in question imply that their discriminants are

$$\Delta_1 = a^2 - 4(3 - b) > 0,$$
$$\Delta_2 = (6 - a)^2 - 4(6 - b) = 0,$$
$$\Delta_3 = (4 - a)^2 - 4(5 - b) < 0,$$

respectively, namely,

$$a^2 + 4b - 12 > 0, \tag{24.2}$$
$$a^2 - 12a + 12 + 4b = 0, \tag{24.3}$$
$$a^2 - 8a - 4 + 4b < 0. \tag{24.4}$$

(24.3) yields

$$a^2 + 4b = 12a - 12. \tag{24.5}$$

Substituting (24.5) into (24.2) yields $12a - 12 - 12 > 0$, i.e. $a > 2$. Substituting (24.5) into (24.4), then $12a - 12 - 8a - 4 < 0$, i.e. $a < 4$. Thus,

$$2 < a < 4. \tag{24.6}$$

(24.3) gives $4b = 24 - (6 - a)^2$. Applying (24.6), it follows that

$$24 - (6 - 2)^2 < 4b < 24 - (6 - 4)^2,$$

so $8 < 4b < 20$ i.e. $2 < b < 5$. Thus, the answer is (A).

Quadratic equation can be applied to solve geometric problems, as shown in the following examples

**Example 9.** $a, b, c$ are positive constants such that the equation

$$c^2 x^2 + (a^2 - b^2 - c^2)x + b^2 = 0$$

has no real roots. Prove that the three segments with lengths $a, b, c$ can form a triangle.

**Solution** The given equation has no real roots implies that

$$(a^2 - b^2 - c^2)^2 - 4b^2 c^2 < 0,$$
$$(a^2 - b^2 - c^2 - 2bc)(a^2 - b^2 - c^2 + 2bc) < 0,$$
$$\therefore [a^2 - (b + c)^2][a^2 - (b - c)^2] < 0.$$

Since $a^2 - (b + c)^2 < a^2 - (b - c)^2$, so $a^2 - (b + c)^2 < 0, a^2 > (b - c)^2$. Thus

$$a < b + c, \quad a > b - c, \quad a > c - b,$$

i.e. $a < b + c, b < a + c, c < a + b$. The conclusion is proven.

In problems about quadratic equations, the relation between roots of two quadratic equations is discussed often, as shown in the following example.

**Example 10.** (CHNMOL/2004) Given that the equation in $x$

$$mx^2 - 2(m + 2)x + m + 5 = 0 \tag{24.7}$$

has no real root, how about the real roots of the following equation

$$(m - 6)x^2 - 2(m + 2)x + m + 5 = 0? \tag{24.8}$$

**Solution**   The equation (24.7) has no real root implies $m \neq 0$ and its discriminant is negative, so

$$[-2(m + 2)]^2 - 4m(m + 5) = -4m + 16 < 0, \quad \text{i.e. } m > 4.$$

For the equation (24.8),

(i)   When $m = 6$, (24.8) becomes $-16x + 11 = 0$, its solution is $x = \dfrac{11}{16}$.

(ii)   When $m \neq 6$, then (24.8) is a quadratic equation, and its discriminant is given by

$$4(m + 2)^2 - 4(m - 6)(m + 5) = 4(5m + 34) > 0 \ (\because m > 4),$$

so (24.8) has two distinct real roots for this case.

Thus, (24.8) has one root $x = \dfrac{11}{16}$ when $m = 6$, or two distinct real roots when $m \neq 6$.

## Testing Questions   (A)

1.   (CHINA/2004) If the larger root of $(2003x)^2 - 2002 \times 2004x - 1 = 0$ is $m$, and the smaller root of $x^2 + 2002x - 2003 = 0$ is $n$, then $m - n$ is

(A) 2004,   (B) 2003,   (C) $\dfrac{2003}{2004}$,   (d) $\dfrac{2002}{2003}$.

2.  (CHINA/2003) Solve the quadratic equation $x^2 + |x + 3| + |x - 3| - 24 = 0$.

3.  (CHINA/2005) Solve the quadratic equation $(m-2)x^2 - (m+3)x - 2m - 1 = 0$.

4.  Given that $a$ is a root of the equation $x^2 - 3x + 1 = 0$, evaluate

$$\frac{2a^5 - 5a^4 + 2a^3 - 8a^2}{a^2 + 1}.$$

5.  (CMO/1988) For what values of $b$ do the equations: $1988x^2 + bx + 8891 = 0$ and $8891x^2 + bx + 1988 = 0$ have a common root?

6.  (CHINA/2004) Given that the equation in $x$

$$(m^2 - 1)x^2 - 2(m + 2)x + 1 = 0$$

has at least a real root, find the range of $m$.

7.  Find the value of $k$, such that the equations $x^2 - kx - 7 = 0$ and $x^2 - 6x - (k + 1) = 0$ have a common root, and find the common root and different roots.

8.  (CHINA/1995) Given that $a, b, c > 0$, and the quadratic equation $(c+a)x^2 + 2bx + (c-a) = 0$ has two equal real roots. Determine if the three segments of lengths $a, b, c$ can form a triangle. If so, what is the type of the triangle? Give your reasons.

9.  If the equation in $x$

$$x^2 + 2(1 + a)x + (3a^2 + 4ab + 4b^2 + 2) = 0$$

has real roots, find the values of $a$ and $b$.

10. (CHINA/1997) $a, b, c$ are real numbers with $a^2 + b^2 + c^2 > 0$. Then the equation

$$x^2 + (a + b + c)x + (a^2 + b^2 + c^2) = 0$$

has

(A) 2 negative real roots,  (B) 2 positive real roots,

(C) 2 real roots with opposite signs,  (D) no real roots.

## Testing Questions    (B)

1. (ASUMO/1990) Mr. Fat is going to pick three non-zero real numbers and Mr. Taf is going to arrange the three numbers as the coefficients of a quadratic equation

$$\Box x^2 + \Box x + \Box = 0.$$

Mr. Fat wins the game if and only if the resulting equation has two distinct rational solutions. Who has a winning strategy?

2. (CHINA/2004) $a, b, c$ are three distinct non-zero real numbers. Prove that the following three equations $ax^2 + 2bx + c = 0, bx^2 + 2cx + a = 0$, and $cx^2 + 2ax + b = 0$ cannot all have two equal real roots.

3. (CHNMOL/2003) $a, b$ are two different positive integers, and the two quadratic equations

$$(a-1)x^2-(a^2+2)x+(a^2+2a) = 0 \quad \text{and} \quad (b-1)x^2-(b^2+2)x+(b^2+2b) = 0$$

have one common root. Find the value of $\dfrac{a^b + b^a}{a^{-b} + b^{-a}}$.

4. (CANADA) $m$ is a real number. Solve the equation in $x$

$$|x^2 - 1| + |x^2 - 4| = mx.$$

5. (CHINA/1988) If $p, q_1$ and $q_2$ are real numbers with $p = q_1 + q_2 + 1$, prove that at least one of the following two equations

$$x^2 + x + q_1 = 0, \qquad x^2 + px + q_2 = 0$$

has two distinct real roots.

# Lecture 25

# Relation between Roots and Coefficients of Quadratic Equations

Viete Theorem and Newton Identity are two important results in discussing the relation between roots and coefficients of polynomial equations. As the fundamental knowledge about the quadratic equation, in this chapter we only mention Viete theorem and its some applications.

**Theorem I.** *(Viete Theorem) If $x_1$ and $x_2$ are the real roots of the equation $ax^2 + bx + c = 0$ $(a \neq 0)$, then*

$$\begin{cases} x_1 + x_2 & = -\dfrac{b}{a}, \\ x_1 x_2 & = \dfrac{c}{a}. \end{cases}$$

**Proof.** By the factor theorem, the equation $ax^2 + bx + c = 0$ $(a \neq 0)$ has roots $x_1$ and $x_2$ if and only if

$$ax^2 + bx + c = a(x - x_1)(x - x_2), \qquad \forall x \in \mathbb{R}.$$

Expanding the right hand side yields $ax^2 - a(x_1 + x_2)x + ax_1x_2$, so the comparison of coefficients of both sides gives the conclusion at once.

Or, the conclusion can be verified by applying the formula for roots. ☐

**Note:** The method used here can be used also for proving the generalized Viete Theorem for polynomial equation of degree $n$ $(n \geq 2)$.

**Theorem II.** *(Inverse Theorem) For any two real numbers $\alpha$ and $\beta$, the equation*

$$x^2 - (\alpha + \beta)x + \alpha\beta = 0$$

*has $\alpha$ and $\beta$ as its two real roots.*

The proof is obvious since the left hand side of the given equation can be factorized to the form

$$(x - \alpha)(x - \beta).$$

When the quadratic equation $ax^2 + bx + c = 0$ is given, i.e. $a, b, c$ are given, by Viete Theorem, not only the roots $x_1$ and $x_2$ can be investigated, but many expressions in $x_1$ and $x_2$ can be given by $a, b, c$, provided the expression is a function of $x_1 + x_2$ and $x_1 x_2$, for example, the following expressions are often used:

$$x_1^2 + x_2^2 = (x_1 + x_2)^2 - 2x_1 x_2 = \frac{b^2}{a^2} - \frac{2c}{a};$$

$$(x_1 - x_2)^2 = (x_1 + x_2)^2 - 4x_1 x_2 = \frac{\Delta}{a^2};$$

$$\frac{1}{x_1} + \frac{1}{x_2} = \frac{x_1 + x_2}{x_1 x_2} = -\frac{b/a}{c/a} = -\frac{b}{c};$$

$$\frac{1}{x_1^2} + \frac{1}{x_2^2} = \frac{x_1^2 + x_2^2}{x_1^2 x_2^2} = \frac{(x_1 + x_2)^2 - 2x_1 x_2}{(x_1 x_2)^2} = \frac{b^2 - 2ac}{c^2};$$

$$x_1^3 + x_2^3 = (x_1 + x_2)(x_1^2 - x_1 x_2 + x_2^2)$$

$$= (x_1 + x_2)[(x_1 + x_2)^2 - 3x_1 x_2] = \left(-\frac{b}{a}\right)^3 + 3\frac{bc}{a^2}.$$

Based on the above, it is possible to establish new equations with required roots by the inverse Viete theorem.

Conversely, by applying the Viete Theorem, the given information on roots of a quadratic equation with parameters can be used to determine the values or ranges of the parameters. To construct quadratic equations by using the inverse Viete theorem can be also used to determine the values of some algebraic expressions. The examples below will explain these applications.

## Examples

**Example 1.** (CHINA/1997) Given that the equation $x^2 + (2a - 1)x + a^2 = 0$ has two real positive roots, where $a$ is an integer. If $x_1$ and $x_2$ are the roots, find the value of $|\sqrt{x_1} - \sqrt{x_2}|$.

**Solution**   The two real roots are positive implies $1 - 2a \geq 0$ i.e. $a \leq \frac{1}{2}$. since $a$ is an integer, so $a \leq 0$. Then

$$|\sqrt{x_1} - \sqrt{x_2}| = \sqrt{(\sqrt{x_1} - \sqrt{x_2})^2} = \sqrt{x_1 + x_2 - 2\sqrt{x_1 x_2}}$$

$$= \sqrt{1 - 2a + 2a} = 1.$$

**Note:** It is not needed to determine the value of $a$ but it is sufficient to know that $a \le 0$.

**Example 2.** (CHNMOL/1996) $x_1$ and $x_2$ are roots of the equation $x^2 + x - 3 = 0$. Find the value of $x_1^3 - 4x_2^2 + 19$.

**Solution** By Viete Theorem, $x_1 + x_2 = -1, x_1 x_2 = -3$. Let $A = x_1^3 - 4x_2^2 + 19$, $B = x_2^3 - 4x_1^2 + 19$. Then

$$
\begin{aligned}
A + B &= (x_1^3 + x_2^3) - 4(x_1^2 + x_2^2) + 38 \\
&= (x_1 + x_2)[(x_1 + x_2)^2 - 3x_1 x_2] - 4[(x_1 + x_2)^2 - 2x_1 x_2] + 38 \\
&= -[1 + 9] - 4[1 + 6] + 38 = 0, \\
A - B &= (x_1^3 - x_2^3) + 4(x_1^2 - x_2^2) \\
&= (x_1 - x_2)[(x_1 + x_2)^2 - x_1 x_2 + 4(x_1 + x_2)] \\
&= (x_1 - x_2)[1 + 3 - 4] = 0.
\end{aligned}
$$

Thus, $2A = (A + B) + (A - B) = 0$, i.e. $A = 0$.

**Example 3.** (CHINA/1996) Given that the quadratic equation $x^2 - px + q = 0$ has two real roots $\alpha$ and $\beta$.
(i)     Find the quadratic equation that takes $\alpha^3, \beta^3$ as its two roots;
(ii)    If the new equation is still $x^2 - px + q = 0$, find all the possible pairs $(p, q)$.

**Solution** (i)   The Viete Theorem yields $\alpha + \beta = p, \alpha\beta = q$, so

$$
\begin{aligned}
(\alpha)^3 \cdot (\beta)^3 &= (\alpha\beta)^3 = q^3, \\
\alpha^3 + \beta^3 &= (\alpha + \beta)^3 - 3\alpha\beta(\alpha + \beta) = p^3 - 3pq = p(p^2 - 3q),
\end{aligned}
$$

by the inverse Viete Theorem, the new equation is

$$
x^2 - p(p^2 - 3q)x + q^3 = 0.
$$

(ii)   If the new equation is the same as the original equation, then

$$
p(p^2 - 3q) = p, \tag{25.1}
$$

$$
q^3 = q. \tag{25.2}
$$

(25.2) implies $q = 0, 1$ or $-1$.
When $q = 0$, then (25.1) implies $p^3 = p$, so $p = 0, 1$ or $-1$.
When $q = 1$, then (25.1) implies $p^3 = 4p$, so $p = 0, 2$ or $-2$.
When $q = -1$, then (25.1) implies $p^3 = -2p$, so $p = 0$.
Corresponding to the obtained seven pairs of $(p, q)$ there are seven equations

$$
\begin{aligned}
& x^2 = 0, \quad x^2 - x = 0, \quad x^2 + x = 0, \quad x^2 + 1 = 0, \\
& x^2 - 2x + 1 = 0, \quad x^2 + 2x + 1 = 0, x^2 - 1 = 0.
\end{aligned}
$$

Among them only $x^2 + 1 = 0$ has no real roots, so the rest 6 pairs satisfy the requirements. Thus, the solutions for $(p, q)$ are

$$(0, 0), \ (1, 0), \ (-1, 0), \ (2, 1), \ (-2, 1), \ (0, -1).$$

**Example 4.** Given the equation $(x - a)(x - a - b) = 1$, where $a, b$ are two constants. Prove that the equation has two real roots, where one is greater than $a$, and the other is less than $a$.

**Solution**　The two roots should be compared with $a$, it is natural to use the substitution $y = x - a$. Under the substitution, the problem becomes one to prove one root is positive and the other is negative for $y$.

Then the equation becomes $y^2 - by - 1 = 0$, and the conclusion is obvious: By Viete theorem, the product of its two roots is $-1$, so they must have opposite signs.

(In fact, $\Delta = b^2 + 4 > 0$ implies that the equation in $y$ must have two distinct real roots, so the existence of real roots is not a problem.)

**Example 5.** (CHNMOL/2000) Given that $m$ is a real number not less than $-1$, such that the equation in $x$

$$x^2 + 2(m - 2)x + m^2 - 3m + 3 = 0$$

has two distinct real roots $x_1$ and $x_2$.

(i)　If $x_1^2 + x_2^2 = 6$, find the value of $m$.

(ii)　Find the maximum value of $\dfrac{mx_1^2}{1 - x_1} + \dfrac{mx_2^2}{1 - x_2}$.

**Solution**　The equation has two distinct real roots implies that $\Delta > 0$, so

$$\Delta = 4(m - 2)^2 - 4(m^2 - 3m + 3) = -4m + 4 > 0,$$

Thus, $-1 \leq m < 1$.

(i)　By Viete Theorem, $x_1 + x_2 = -2(m - 2)$, $x_1 x_2 = m^2 - 3m + 3$, so

$$\begin{aligned}
6 &= x_1^2 + x_2^2 = (x_1 + x_2)^2 - 2x_1 x_2 \\
&= 4(m - 2)^2 - 2(m^2 - 3m + 3) = 2m^2 - 10m + 10,
\end{aligned}$$

i.e. $m^2 - 5m + 2 = 0$, so $m = \dfrac{5 \pm \sqrt{17}}{2}$. Since $-1 \leq m < 1$,

$$m = \frac{5 - \sqrt{17}}{2}.$$

(ii) $\dfrac{mx_1^2}{1-x_1} + \dfrac{mx_2^2}{1-x_2} = \dfrac{m[x_1^2(1-x_2) + x_2^2(1-x_1)]}{(1-x_1)(1-x_2)}$

$= \dfrac{m[x_1^2 + x_2^2 - x_1x_2(x_1+x_2)]}{x_1x_2 - (x_1+x_2) + 1}$

$= \dfrac{m[(2m^2 - 10m + 10) + 2(m^2 - 3m + 3)(m-2)]}{m^2 - 3m + 3 + 2(m-2) + 1}$

$= \dfrac{m(2m^3 - 8m^2 + 8m - 2)}{m^2 - m} = 2(m^2 - 3m + 1)$

$= 2\left(m - \dfrac{3}{2}\right)^2 - \dfrac{5}{2} \le 2\left(-1 - \dfrac{3}{2}\right)^2 - \dfrac{5}{2} = 10 \quad$ since $-1 \le m < 1$.

Thus, the maximum value of $\dfrac{mx_1^2}{1-x_1} + \dfrac{mx_2^2}{1-x_2}$ is 10.

**Example 6.** (CHNMOL/1991) Given that the quadratic equation $ax^2 + bx + c = 0$ has no real roots, but Adam got two roots 2 and 4 since he wrote down a wrong value of $a$. Ben also got two roots $-1$ and 4 because he wrote down the sign of a term wrongly. Find the value of $\dfrac{2b + 3c}{a}$.

**Solution** Suppose that Adam wrote down the coefficient $a$ as $a'$ wrongly. Then

$$-\frac{b}{a'} = 6, \quad \frac{c}{a'} = 8,$$

therefore $-\dfrac{b}{c} = \dfrac{3}{4}$. Based on Ben's result, we have

$$\left|\frac{b}{a}\right| = 3, \quad \left|\frac{c}{a}\right| = 4.$$

Since $\Delta = b^2 - 4ac < 0$, we have $ac > 0$, therefore $\dfrac{c}{a} = 4$, and hence $\dfrac{b}{a} = -3$. Thus,

$$\frac{2b + 3c}{a} = 2(-3) + 3(4) = 6.$$

**Example 7.** (CHINA/2003) Given that the equation $8x^2 + (m+1)x + m - 7 = 0$ has two negative roots, find the range of the parameter $m$.

**Solution** Let the roots be $x_1, x_2$. Then $x_1 + x_2 = -\dfrac{m+1}{8} < 0$, $x_1x_2 = \dfrac{m-7}{8} > 0$ and $\Delta = (m+1)^2 - 32(m-7) \ge 0$.

The first inequality implies that $-1 < m$, the second inequality implies that $7 < m$, and

$$(m+1)^2 - 32(m-7) = m^2 - 30m + 225 = (m-15)^2 \geq 0,$$

which implies that $m$ can be any real value. Thus, the range of $m$ is $7 < m$.

**Example 8.** (CHINA/2005) If $a, b$ are real numbers and $a^2 + 3a + 1 = 0, b^2 + 3b + 1 = 0$, find the values of $\dfrac{a}{b} + \dfrac{b}{a}$.

**Solution**   From Viete theorem it follows that $a$ and $b$ are both the real roots of the equation $x^2 + 3x + 1 = 0$. If $a = b$ then

$$\frac{a}{b} + \frac{b}{a} = 2.$$

If $a \neq b$, then Viete Theorem yields that $a + b = -3, ab = 1$, so

$$\frac{a}{b} + \frac{b}{a} = \frac{(a+b)^2 - 2ab}{ab} = 9 - 2 = 7.$$

**Example 9.** (CHINA/2005) If $p, q$ are two real numbers satisfying the relations $2p^2 - 3p - 1 = 0$ and $q^2 + 3q - 2 = 0$ and $pq \neq 1$. Find the value of $\dfrac{pq + p + 1}{q}$.

**Solution**   Change the second equality to the form $2\left(\dfrac{1}{q}\right)^2 - 3\left(\dfrac{1}{q}\right) - 1 = 0$, then it is found that $p$ and $1/q$ are both the roots of the equation $2x^2 - 3x - 1 = 0$. $pq \neq 1$ implies that $p \neq 1/q$, so, by Viete theorem,

$$p + \frac{1}{q} = \frac{3}{2} \qquad \text{and} \qquad \frac{p}{q} = -\frac{1}{2},$$

hence $\dfrac{pq + p + 1}{q} = p + \dfrac{1}{q} + \dfrac{p}{q} = \dfrac{3}{2} - \dfrac{1}{2} = 1.$

**Example 10.** Given that $a, b, c$ are the lengths of three sides of $\triangle ABC, a > b > c$, $2b = a + c$, and $b$ is a positive integer. If $a^2 + b^2 + c^2 = 84$, find the value of $b$.

**Solution**   The conditions $a + c = 2b$ and $a^2 + b^2 + c^2 = 84$ yield

$$ac = \frac{1}{2}[(a+c)^2 - (a^2 + c^2)] = \frac{1}{2}(5b^2 - 84).$$

By the inverse Viete theorem, the equation $x^2 - 2bx + \dfrac{5b^2 - 84}{2} = 0$ has two distinct roots $a$ and $c$, so its discriminant is positive, i.e. $\Delta = 4b^2 - 2(5b^2 - 84) = 168 - 6b^2 > 0$, which implies

$$b^2 < \frac{168}{6} = 28.$$

Since $ac > 0$ implies $84 < 5b^2$, so $16 < 84/5 < b^2$. Thus, $16 < b^2 < 28$ yields $b = 5$.

## Testing Questions (A)

1. (CHINA/1999) Given that $2x^2 - 5x - a = 0$ is a quadratic equation in $x$, $a$ is a parameter. If the ratio of its two roots $x_1 : x_2 = 2 : 3$, find the value of $x_2 - x_1$.

2. (CHINA/1993) Given that the two roots of the equation $x^2 + px + q = 0$ are 1 greater than the two roots of the equation $x^2 + 2qx + \frac{1}{2}p = 0$ respectively, and the difference of the two roots of $x^2 + px + q = 0$ is equal to that of $x^2 + 2qx + \frac{1}{2}p = 0$. Find the solutions to each of the two equations.

3. (CHINA/1993) $\alpha, \beta$ are the real roots of the equation $x^2 - px + q = 0$. Find the number of the pairs $(p, q)$ such that the quadratic equation with roots $\alpha^2, \beta^2$ is still $x^2 - px + q = 0$.

4. (RUSMO/1989) Given $p + q = 198$, find the integer solutions of the equation $x^2 + px + q = 0$.

5. (CHINA/1995) Given that the sum of squares of roots to the equation $2x^2 + ax - 2a + 1 = 0$ is $7\frac{1}{4}$, find the value of $a$.

6. Given that $\alpha$ and $\beta$ are the real roots of $x^2 - 2x - 1 = 0$, find the value of $5\alpha^4 + 12\beta^3$.

7. (CHINA/1997) Given that $\alpha$ and $\beta$ are the real roots of $x^2 + 19x - 97 = 0$, and $\dfrac{1 + \alpha}{1 - \alpha} + \dfrac{1 + \beta}{1 - \beta} = -\dfrac{m}{n}$, where $m$ and $n$ are two relatively prime natural numbers. Find the value of $m + n$.

8. (CHINA/1997) Given that $a, b$ are integers with $a > b$, and the two roots $\alpha, \beta$ of the equation $3x^2 + 3(a + b)x + 4ab = 0$ satisfy the relation

$$\alpha(\alpha + 1) + \beta(\beta + 1) = (\alpha + 1)(\beta + 1),$$

find all the pairs $(a, b)$ of two integers.

9.  (CHNMOL/1999) Given that the real numbers $s, t$ satisfy $19s^2 + 99s + 1 = 0$, $t^2 + 99t + 19 = 0$, and $st \neq 1$. Find the value of $\dfrac{st + 4s + 1}{t}$.

10.  (USSR) Prove that if $\alpha$ and $\beta$ are the roots of the equation $x^2 + px + 1 = 0$, and if $\gamma$ and $\delta$ are the roots of the equation $x^2 + qx + 1 = 0$, then

$$(\alpha - \gamma)(\beta - \gamma)(\alpha + \delta)(\beta + \delta) = q^2 - p^2.$$

# Testing Questions　　(B)

1.  (CHINA/1998) Given that $\alpha, \beta$ are roots of the equation $x^2 - 7x + 8 = 0$, where $\alpha > \beta$. Find the value of $\dfrac{2}{\alpha} + 3\beta^2$ without solving the equation.

2.  Given that $a = 8 - b$ and $c^2 = ab - 16$, prove that $a = b$.

3.  (USSR) Let $\alpha$ and $\beta$ be the roots of the equation $x^2 + px + q = 0$, and $\gamma$ and $\delta$ be the roots of the equation $x^2 + Px + Q = 0$. Express the product

$$(\alpha - \gamma)(\beta - \gamma)(\alpha - \delta)(\beta - \delta)$$

in terms of the coefficients of the given equations.

4.  (ASUMO/1986) If the roots of the quadratic equation $x^2 + ax + b + 1 = 0$ are natural numbers, prove that $a^2 + b^2$ is a composite number.

5.  (CHINA/1999) Solve the equation $\dfrac{13x - x^2}{x + 1}\left(x + \dfrac{13 - x}{x + 1}\right) = 42$.

# Lecture 26

# Diophantine Equations (II)

The Diophantine equations to be discussed in this chapter are all non-linear although, but we focus on the integer solutions of quadratic equations.

## Basic Methods for Solving Quadratic Equations on $\mathbb{Z}$

(I)   Factorization Method.   Let the right hand side of the equation be a constant, zero or a power of a prime number (in the case of indicial equation), and factorize the left hand side to the form of product of linear factors, then discuss the possible values of the linear factors based on the factorization of the right hand side.

(II)  Discriminant Method.   When a quadratic equation with integer coefficients has integer solution(s), its discriminant must be a perfect square. This feature will play an important role.
     When the quadratic equation contains two variables $x, y$, by the formula for roots, $x$ can be expressed in terms of $y$, and its discriminant is an expression in $y$. Since the discriminant is a perfect square and that $y$ is an integer, $y$ can be found easily in many cases.

     Besides the use of discriminant, the use of Viete Theorem and transformation or substitution are also useful tools for simplifying and solving quadratic equations.

(III) Congruence, divisibility, and parity analysis, etc. are often used in discussing the existence of integer solutions of quadratic equations.

69

## Examples

**Example 1.** (SSSMO(J)/2008) Let $n$ be a positive integer such that $n^2 + 19n + 48$ is a perfect square. Find the value of $n$.

**Solution** This question can be solved by factorization method.
Let $n^2 + 19n + 48 = m^2$ for some $m \in \mathbb{N}$, then $4n^2 + 76n + 192 = 4m^2$, so

$$(2n + 19)^2 - (2m)^2 = 169.$$
$$(2n - 2m + 19)(2n + 2m + 19) = 13^2 = 1 \cdot 169 = 13 \cdot 13.$$

The difference of $2n - 2m + 19 = 1$ and $2n + 2m + 19 = 169$ yields $4m = 168$ i.e. $m = 42$. Therefore $n = m - 9 = 33$. By checking, 33 satisfies the given equation.

The difference of $2n - 2m + 19 = 13$ and $2n + 2m + 19 = 13$ yields $m = 0$, its impossible for $n > 0$. So $n = 33$ is the unique required solution.

**Example 2.** (CHINA/2003) Find the integer solutions of the equation $6xy + 4x - 9y - 7 = 0$.

**Solution** By factorization, $6xy + 4x - 9y - 6 = (2x - 3)(3y + 2)$, so the given equation becomes

$$(2x - 3)(3y + 2) = 1.$$

If $2x - 3 = 1, 3y + 2 = 1$, then $y$ has no integer solution.
If $2x - 3 = -1, 3y + 2 = -1$, then $x = 1, y = -1$. By checking, $(1, -1)$ satisfies the original equation, so it is the unique solution for $(x, y)$.

**Example 3.** (SSSMO(J)/2004) Find the number of ordered pairs of positive integers $(x, y)$ that satisfy the equation

$$\frac{1}{x} + \frac{1}{y} = \frac{1}{2004}.$$

**Solution** By removing the denominators, it follows that $xy = 2004(x + y)$. Then $xy - 2004x - 2004y + 2004^2 = 2004^2$, so

$$(x - 2004)(y - 2004) = 2004^2.$$

For any positive factor $p$ of $2004^2$, $2004^2 = p \cdot (2004^2/p)$ yields a solution $x = 2004 + p, y = 2004 + (2004^2/p)$. Since

$$2004^2 = (2^2 \times 3 \times 167)^2 = 2^4 \times 3^2 \times 167^2,$$

the number of positive divisors of $2004^2$ is $(4 + 1)(2 + 1)(2 + 1) = 45$, so there are 45 such solutions.

If $p$ is a negative factor of $2004^2$, then so is $2004^2/p$, and one of them must have absolute value not less than 2004, so one of $p + 2004$ and $(2004^2/p) + 2004$ is not positive, i.e., such solutions are not required. Thus, there are a total of 45 required solutions.

**Example 4.** (SSSMO(J)/2009) Find the value of the smallest positive integer $m$ such that the equation

$$x^2 + 2(m + 5)x + (100m + 9) = 0$$

has only integer solutions.

**Solution** Since the equation has integer solutions,

$$\Delta = 4[(m + 5)^2 - (100m + 9)] = 4(m^2 - 90m + 16) \geq 0$$

and $m^2 - 90m + 16 = n^2$ for some non-negative integer $n$. Then $(m - 45)^2 + 16 - 2025 = n^2$, so

$$(m - n - 45)(m + n - 45) = 2009$$
$$= 1 \cdot 2009 = 7 \cdot 287 = 41 \cdot 49 = (-49)(-41) = (-287)(-7) = (-2009)(-1).$$

By solving the six corresponding systems of two equations in $m, n$ we have

$$(m, n) = (1050, 1004), (192, 140), (90, 4), (-102, 140), (0, 4), (-960, 1004).$$

Since $x = -(m + 5) \pm n$, all above values of $(m, n)$ give integral roots. Thus, the smallest positive value of $m$ is 90.

Viete Theorem is also used for solving Diophantine equations. Below is an example.

**Example 5.** (CHNMOL/2005) $p, q$ are two integers, and the two roots of the equation in $x$

$$x^2 - \frac{p^2 + 11}{9}x + \frac{15}{4}(p + q) + 16 = 0$$

are $p$ and $q$ also. Find the values of $p$ and $q$.

**Solution** Viete Theorem yields

$$p + q = \frac{p^2 + 11}{9}, \tag{26.1}$$

$$pq = \frac{15}{4}(p + q) + 16. \tag{26.2}$$

Then $p + q > 0$ from (26.1) and $pq > 0$ from (26.2), so $p, q$ are both positive integers. From (26.2) it follows that

$$16pq - 60(p + q) = 16^2$$
$$\therefore (4p - 15)(4q - 15) = 256 + 225 = 481.$$

Since $481 = 1 \times 481 = 13 \times 37 = (-1) \times (-481) = (-13) \times (-37)$, and $4p - 15$ or $4q - 15$ cannot be $-37$ or $-481$, so the pair $(4p - 15, 4q - 15)$ has the following four possible cases:

$$(1, 481), \quad (481, 1), \quad (13, 37), \quad (37, 13).$$

Corresponding to them, the pairs of $(p, q)$ are

$$(4, 124), \quad (124, 4), \quad (7, 13), \quad (13, 7).$$

By checking, only the pair $(13, 7)$ satisfies the original system: the equation becomes

$$x^2 - 20x + 91 = 0,$$

and its roots are $\{13, 7\}$. Thus, the solution for $(p, q)$ is $(13, 7)$.

Many techniques often used in number theory, like congruence, divisibility, parity analysis, etc., can be used for solving Diophantine equations. Below a few such examples are given.

**Example 6.** (KIEV/1962) Prove that the equation $x^2 + y^2 = 3z^2$ has no integer solution $(x, y, z) \neq (0, 0, 0)$.

**Solution**   First of all it can be shown that, if an integer solution $(x, y, z)$ is not $(0, 0, 0)$, then there must be such an integer solution with $(x, y) = 1$.

Suppose that $(x, y, z) \neq (0, 0, 0)$ is an integer solution with $(x, y) = d > 1$, letting

$$x = dx_1, \quad y = dy_1, \quad \text{with } (x_1, y_1) = 1,$$

then the original equation becomes $d^2(x_1^2 + y_1^2) = 3z^2$, so $d^2 \mid 3z^2$. Since the indices of 3 in $d^2$ and $z^2$ are both even, so $d^2 \mid z^2$, i.e. $d \mid z$. Let $z = dz_1$, then $x_1^2 + y_1^2 = 3z_1^2$. Thus, $(x_1, y_1, z_1) \neq (0, 0, 0)$ is an integer solution of the given equation also, and $x_1, y_1, z_1$ are relatively prime pairwise.

Hence, it suffices to show that the given equation has no non-zero integer solutions $(x, y, z)$ with $(x, y) = 1$.

Suppose that $(x, y, z)$ is such a solution, then $x, y$ cannot be divisible by 3, and $x^2 + y^2 \equiv 2 \pmod 3$ if $x, y$ are both not divisible by 3, a contradiction.

Thus, the conclusion is proven.

**Example 7.** (USAMO/1975) Determine all integral solutions of

$$a^2 + b^2 + c^2 = a^2 b^2.$$

**Solution**  Let $(a, b, c)$ be an integer solution. We show that $a, b$ and $c$ must all be even by taking modulo 4 to both sides of the equation. There are three possible cases to be considered:

*Case 1*:  When $a, b, c$ are all odd, then $a^2 + b^2 + c^2 \equiv 3$ (mod 4), whereas $a^2 b^2 \equiv 1$ (mod 4), so it is impossible.

*Case 2*:  When two of $a, b, c$ are odd and the other is even, then $a^2 + b^2 + c^2 \equiv 2$ (mod 4), whereas $a^2 b^2 \equiv 0$ or 1 (mod 4), so it is impossible also.

*Case 3*:  When two of $a, b, c$ are even and the other is odd, then $a^2 + b^2 + c^2 \equiv 1$ (mod 4), whereas $a^2 b^2 \equiv 0$ (mod 4), so it is impossible also.

Thus, $a, b, c$ are all even. Let $a = 2a_1, b = 2b_1, c = 2c_1$, this leads to the relation

$$a_1^2 + b_1^2 + c_1^2 = 4a_1^2 b_1^2.$$

Since $4a_1^2 b_1^2 \equiv 0$ (mod 4) and each of $a_1^2, b_1^2, c_1^2$ has remainder 0 or 1 modulo 4, $a_1, b_1, c_1$ must all be even also. Then $a_1 = 2a_2, b_1 = 2b_2, c_1 = 2c_2$. This leads to the relation

$$a_2^2 + b_2^2 + c_2^2 = 16a_2^2 b_2^2.$$

Again we can conclude that $a_2, b_2, c_2$ are all even, so it leads to the relation

$$a_3^2 + b_3^2 + c_3^2 = 64a_3^2 b_3^2.$$

The process can be continued to any times, since we have the relation

$$a_n^2 + b_n^2 + c_n^2 = 4^n a_n^2 b_n^2$$

for any natural number $n$. Hence $a_n = \dfrac{a}{2^n}, b_n = \dfrac{b}{2^n}, c_n = \dfrac{c}{2^n}$ are integers for any natural number $n$, i.e. $a = b = c = 0$.

Thus, the equation has only zero solution.

**Note:**  This example is one of Fermat's method of **infinite decent**.

It is sometimes useful to use substitution to simplify the equation first, below is such an example.

**Example 8.** (CHNMOL/2003) Given that the integers $a, b$ satisfy the equation

$$\left[ \frac{\dfrac{1}{a} - \dfrac{1}{b}}{\dfrac{1}{a} - \dfrac{1}{b}} \quad \frac{1}{\dfrac{1}{a} + \dfrac{1}{b}} \right] \left( \frac{1}{a} - \frac{1}{b} \right) \cdot \frac{1}{\dfrac{1}{a^2} + \dfrac{1}{b^2}} = \frac{2}{3},$$

find the value of $a + b$.

**Solution**   Let $x = \dfrac{1}{a}, y = \dfrac{1}{b}$, then the left hand side of the given equation becomes

$$\left[ \frac{x}{x-y} - \frac{y}{x+y} \right] (x-y) \cdot \frac{1}{x^2 + y^2}$$
$$= \frac{x^2 + y^2}{(x-y)(x+y)} \cdot (x-y) \cdot \frac{1}{x^2 + y^2} = \frac{1}{x+y}.$$

Then the given equation is simplified to $\dfrac{1}{x+y} = \dfrac{2}{3}$, i.e.

$$\frac{1}{\frac{1}{a} + \frac{1}{b}} = \frac{2}{3}, \quad \text{or} \quad \frac{ab}{a+b} = \frac{2}{3}.$$

From it we have $3ab - 2a - 2b = 0$ which yields $9ab - 6a - 6b + 4 = 4$, i.e.

$$(3a - 2)(3b - 2) = 4.$$

Therefore $a \neq b$. By symmetry, we may assume $a > b$, so
(i)   when $3a - 2 = 4, 3b - 2 = 1$ then $a = 2, b = 1, a + b = 3$;
(ii)   when $3a - 2 = -1, 3b - 2 = -4$ then $a$ has no integer solution.
Thus, $a + b = 3$.

**Example 9.** (CHNMOL/1995) The number of positive integer solutions $(x, y, z)$ for the system of simultaneous equations

$$\begin{cases} xy + yz = 63, \\ xz + yz = 23 \end{cases}$$

is

(A) 1;       (B) 2;       (C) 3;       (D) 4.

**Solution.**   23 is a prime, so that the second equation has less uncertainty, we deal with it first. $x + y \geq 2$ and 23 is prime leads $z = 1, x + y = 23$. By substituting $y = 23 - x$ into the first equation, it follows that

$$(23 - x)(x + 1) = 63,$$
$$x^2 - 22x + 40 = 0,$$
$$(x - 2)(x - 20) = 0,$$

so $x_1 = 2, x_2 = 20$. Then $y_1 = 21, y_2 = 3$. Thus the solutions are $(2, 21, 1)$ and $(20, 3, 1)$, the answer is (B).

## Testing Questions    (A)

1.  (CHINA/2003) Given that $\dfrac{1260}{a^2 + a - 6}$ is a positive integer, where $a$ is a positive integer. Find the value of $a$.

2.  (CHINA/2001) How many number of pairs $(x, y)$ of two integers satisfy the equation
$$x^2 - y^2 = 12?$$

3.  (SSSMO(J)/2004) Let $x, y, z$ and $w$ represent four *distinct* positive integers such that
$$x^2 - y^2 = z^2 - w^2 = 81.$$
Find the value of $xz + yw + xw + yz$.

4.  (CHINA/2003) Find the number of non-zero integer solutions $(x, y)$ to the equation $\dfrac{15}{x^2 y} + \dfrac{3}{xy} - \dfrac{2}{x} = 2$.

5.  (CHINA/2001) Find the number of positive integer solutions to the equation
$$\frac{x}{3} + \frac{14}{y} = 3.$$

6.  (CHINA/2001) Find the number of positive integer solutions of the equation $\dfrac{2}{x} - \dfrac{3}{y} = \dfrac{1}{4}$.

7.  (SSSMO/2005) How many ordered pairs of integers $(x, y)$ satisfy the equation
$$x^2 + y^2 = 2(x + y) + xy?$$

8.  (SSSMO/2003) Let $p$ be a positive prime number such that the equation
$$x^2 - px - 580p = 0$$
has two integer solutions. Find the value of $p$.

9.  (USSR/1962) Prove that the only solution in integers of the equation
$$x^2 + y^2 + z^2 = 2xyz$$
is $x = y = z = 0$.

10.    (CHINA/1993) The number of positive integer solutions $(x, y, z)$ for the system of simultaneous equations

$$\begin{cases} xy + xz = 255, \\ xy + yz = 31 \end{cases}$$

is

   (A) 3;       (B) 2;       (C) 1;       (D) 0.

# Testing Questions    (B)

1.    (IMO/Shortlist/1989) Given the equation

$$4x^3 + 4x^2y - 15xy^2 - 18y^3 - 12x^2 + 6xy + 36y^2 + 5x - 10y = 0,$$

   find all positive integer solutions.

2.    (USSR) Solve, in integers, $\dfrac{1}{x} + \dfrac{1}{y} = \dfrac{1}{z}$ (find the formula for general solution.)

3.    (SSSMO/2006) Let $p$ be an integer such that both roots of the equation

$$5x^2 - 5px + (66p - 1) = 0$$

   are positive integers. Find the value of $p$.

4.    (RUSMO/1991) Find all the natural numbers $p, q$ such that the equation $x^2 - pqx + p + q = 0$ has two integer roots.

5.    (IMO/1994) Determine all ordered pairs $(m, n)$ of positive integers such that

$$\frac{n^3 + 1}{mn - 1}$$

   is an integer.

# Lecture 27

# Linear Inequality and System of Linear Inequalities

**Definition 1**  An expression is called an **Inequality** if it is formed by two algebraic expressions connected by the symbols ">", "$\geq$", "<", or "$\leq$", aimed to represent the unequal relation between the two expressions.

If use the letters $a$ and $b$ to stand for the two expressions in an inequality, then $a > b$ is called "$a$ is greater than $b$", $a \geq b$ is called "$a$ is greater than or equal to $b$", $a < b$ is called "$a$ is less than $b$", and $a \leq b$ is called "$a$ is less than or equal to $b$".

**Definition 2**  When an inequality (or a system of inequalities) has $k$ unknown variables $(x_1, x_2, \ldots, x_k) \in \mathbb{R}_k$, the action to find the values of $(x_1, x_2, \ldots, x_k)$ such that the inequality (or the system of the inequalities) is (are) true is called **solving** the inequality (or system of the inequalities).  Any point $(x_1, x_2, \ldots, x_k)$ $\in \mathbb{R}_k$ that satisfies the inequality (or the system of inequalities) is called a **solution** of the inequality (or the system of the inequalities), and the set of all solutions is called the **solutions set** of the inequality (or system of inequalities).

## Basic Properties of Inequalities

(I)    If $a > b$ and $b > c$, then $a > c$.

(II)   If $a > b$, then $a + c > b + c$ and $a - c > b - c$ for any real number $c$.

(III)  If $a > b$, then $a \cdot c > b \cdot c$ and $\dfrac{a}{c} > \dfrac{b}{c}$ if $c > 0$.

(IV)   If $a > b$, then $a \cdot c < b \cdot c$ and $\dfrac{a}{c} < \dfrac{b}{c}$ if $c < 0$, i.e. **the direction of the inequality needs to be changed**.

Note that the property (IV) is exclusive for inequalities, but the properties (I) to (III) are similar to the case of equalities.

**Steps for Solving a Linear Inequality**

When an inequality has a unknown variable $a$ to be solved, the steps are usually the same as in solving a linear equation, consisting of

(i) removing denominators;        (ii) removing brackets;

(iii) moving terms for combining like terms;        (iv) combining like terms;

(v) normalizing the coefficient of $x$.

The order of (i) to (iv) can be changed flexibly, such that the inequality can be simplified to one of the forms

$$ax > b, \quad ax \geq b, \quad ax < b, \quad ax \leq b,$$

where $a, b$ are constants. At the step (v), $a$ should be converted to 1 if $a$ is given constant, however, if $a$ is a parameter or contains parameters, then the discussion on the possible cases of the parameters is needed.

**Examples**

**Example 1.** Given $2(x - 2) - 3(4x - 1) = 9(1 - x)$ and $y < x + 9$, compare the sizes of $\dfrac{y}{\pi}$ and $\dfrac{10}{31}y$.

**Solution**   By solving the equation in $x$, it follows that $12x - 2x - 9x = -4 + 3 - 9 = -10$, so $x = -10$. Thus, $y < -10 + 9 = -1$. Since $\dfrac{1}{\pi} < \dfrac{10}{31}$, so $\dfrac{1}{\pi}y > \dfrac{10}{31}y$.

**Example 2.** (CHINA/1999) If $ac < 0$, in the inequalities $\dfrac{a}{c} < 0$; $ac^2 < 0$; $a^2c < 0$; $c^3a < 0$; $ca^3 < 0$ how many must be true?

(A) 1,        (B) 2,        (C) 3,        (D) 4.

**Solution**   $ac < 0$ implies that $a$ and $c$ have opposite signs and $a \neq 0, c \neq 0$, so $\dfrac{a}{c} < 0$ and $a^2 > 0$ and $c^2 > 0$. Since the signs of $a$ and $c$ are not determined, so the signs of $a^2c$ and $ac^2$ are not determined. However, $a^3c = a^2(ac) < 0$ and $ca^3 = (ac)a^2 < 0$, so there are three inequalities that must be true. The answer is (C).

**Example 3.** (CHINA/2006) There are four statements as follows:

(i)   When $0 < x < 1$, then $\dfrac{1}{1 + x} < 1 - x + x^2$;

(ii)   When $0 < x < 1$, then $\dfrac{1}{1+x} > 1 - x + x^2$;

(iii)   When $-1 < x < 0$, then $\dfrac{1}{1+x} < 1 - x + x^2$;

(iv)   When $-1 < x < 0$, then $\dfrac{1}{1+x} > 1 - x + x^2$.

Then the correct statements are

(A) (i) and (iii),   (B) (ii) and (iv),   (C) (i) and (iv),   (D) (ii) and (iii).

**Solution**   Under the conditions on $x$, $1 + x > 0$, so we can move $1 + x$ to right hand side without changing the direction of the inequality sign. Since

$$(1 + x)(1 - x + x^2) = 1 + x^3,$$

so (i) and (iv) are true but (ii) and (iv) are wrong. Thus, the answer is (C).

**Example 4.** (CHINA/2006) Given real numbers $a$ and $b$. If $a = \dfrac{x+3}{4}, b = \dfrac{2x+1}{3}, b < \dfrac{7}{3} < 2a$, find the range of the value of $x$.

**Solution**   The inequalities $b < \dfrac{7}{3} < 2a$ implies $\dfrac{2x+1}{3} < \dfrac{7}{3} < \dfrac{x+3}{2}$, so $2x + 1 < 7$ and $14 < 3x + 9$, hence $\dfrac{5}{3} < x < 3$. Thus, the range of $x$ is $\dfrac{5}{3} < x < 3$.

**Example 5.** (CHINA/2004) If the solution set of the inequality $(a + 1)x > a^2 - 1$ is $x < a - 1$, find the solution set of the inequality $(1 - a)x < a^2 - 2a + 1$.

**Solution**   The given conditions in question implies that $a + 1 < 0$. In fact, $a + 1 < 0$ leads to the solution set of the first inequality is $x < \dfrac{a^2 - 1}{a + 1}$ i.e. $x < a - 1$.

Then $a + 1 < 0$ implies $a < -1$, so $1 - a > 0$. Thus, the solution set of the second inequality is

$$x < \dfrac{a^2 - 2a + 1}{1 - a} = 1 - a, \quad \text{i.e. } x < 1 - a.$$

**Example 6.** (CHINA/1998) Solve the inequality $a(x - b^2) > b(x + a^2)$ for $x$.

**Solution**   $a(x - b^2) > b(x + a^2)$ yields $(a - b)x > ab(a + b)$.

(i)     If $a > b$, then $x > \dfrac{ab(a + b)}{a - b}$;

(ii)    If $a < b$, then $x < \dfrac{ab(a + b)}{a - b}$;

(iii)   If $a = b \geq 0$, then left hand side is zero but right hand side is non-negative, no solution;

(iv)    If $a = b < 0$, then left hand side is zero and right hand side is negative, so any real number is a solution.

**Example 7.** Given that the solution set of the inequality $(2a - b)x + 3a - 4b < 0$ is $x > \dfrac{4}{9}$. Find the solution of the inequality $(a - 4b)x + 2a - 3b > 0$.

**Solution**   The solution set of $(2a - b)x + 3a - 4b < 0$ is $x > \dfrac{4}{9}$ implies that

$$2a - b < 0 \quad \text{and} \quad \frac{4b - 3a}{2a - b} = \frac{4}{9},$$
$$2a < b, \quad \text{and} \quad 36b - 27a = 8a - 4b,$$
$$\therefore b = \frac{7}{8}a > 2a \Rightarrow a < 0.$$

Then $(a - 4b)x + 2a - 3b > 0$ becomes $\left(a - \dfrac{7}{2}a\right)x + 2a - \dfrac{21}{8}a > 0$, so

$$-\frac{5}{2}ax > \frac{5}{8}a, \quad \therefore x > -\frac{1}{4}.$$

Thus, the solution set of the second inequality is $x > -\dfrac{1}{4}$.

**Example 8.** (CHINA/2001) Given that the solution set for $x$ of the inequality $\dfrac{2m + x}{3} \leq \dfrac{4mx - 1}{2}$ is $x \geq \dfrac{3}{4}$, find the value of the parameter $m$.

**Solution**   From $\dfrac{2m + x}{3} \leq \dfrac{4mx - 1}{2}$ it follows that

$$4m + 2x \leq 12mx - 3,$$
$$2(6m - 1)x \geq 4m + 3.$$

The solution set is $x \geq \dfrac{3}{4}$ implies that $6m - 1 > 0$ and $\dfrac{4m + 3}{2(6m - 1)} = \dfrac{3}{4}$,

$$\therefore 2(4m + 3) = 3(6m - 1), \quad \text{i.e. } (18 - 8)m = 6 + 3, \quad \therefore m = \frac{9}{10}.$$

**Example 9.** (CHINA/2005) Given that $x = 3$ is a solution of the inequality $mx + 2 < 1 - 4m$, if $m$ is an integer, find the maximum value of $m$.

**Solution** By substituting $x = 3$ into the given inequality, it follows that $7m < -1$, so $m < -\dfrac{1}{7}$, thus, the maximum value of $m$ is $-1$.

Finding the solution set of a system of linear inequalities, means to find the set which satisfies each of the inequality in the system. So to each inequality in the system we can find its solution set first, then the common part of these sets is the solution set of the system.

**Example 10.** Solve the system of inequalities
$$\begin{cases} 3x - 5 \geq 2x - 3, \\ 2(3x + 2) \geq 3x - 1. \end{cases}$$

**Solution** The solution set of the first inequality is $x \geq 2$.

The solution set of the second inequality is $x \geq -\dfrac{5}{3}$.

Therefore the solution set of the system is the common part of these two intervals, i.e. the set $x \geq 2$.

## Testing Questions    (A)

1. If $a, b, c > 0$ and $\dfrac{c}{a+b} < \dfrac{a}{b+c} < \dfrac{b}{a+c}$, arrange $a, b, c$ in ascending order.

2. (CHINA/2005) Given $a < b < c < 0$, arrange the sizes of $\dfrac{a}{b+c}, \dfrac{b}{c+a}, \dfrac{c}{a+b}$ in descending order.

3. Solve the inequality in $x$: $ax + 4 < x + b$, where $a, b$ are two constants.

4. (CHINA/2002) Given $m = \dfrac{4-x}{3}, n = \dfrac{x+3}{4}, p = \dfrac{2-3x}{5}$, and $m > n > p$. Find the range of $x$.

5. Solve the system of inequalities
$$\begin{cases} x - 1 > -3 \\ \dfrac{1}{2}x - 1 < \dfrac{1}{3}x \\ 3 < 2(x - 1) < 10 \\ \dfrac{1}{3}(3 - 2x) > -2. \end{cases}$$

6.  (CHINA/1996) Given that $x, y, a, b$ are all positive numbers and $a < b, \dfrac{x}{y} = \dfrac{a}{b}$. If $x + y = c$, then the larger one of $x$ and $y$ is

    (A) $\dfrac{ab}{a+b}$,   (B) $\dfrac{ab}{b+c}$,   (C) $\dfrac{ac}{a+b}$,   (D) $\dfrac{bc}{a+b}$.

7.  Given that the solution set of the inequality $(2a - b)x > a - 2b$ for $x$ is $x < \dfrac{5}{2}$, find the solution set of the inequality $ax + b < 0$.

8.  Given $a + b + c = 0, a > b > c$. Find the range of $\dfrac{c}{a}$.

9.  (CHINA/1997) Given that the solution set of $x$ for the inequality $(2a - b)x + a - 5b > 0$ is $x < \dfrac{10}{7}$, find the solution set of $x$ for the inequality $ax > b$.

10. (CHINA/1998) Given that $a, b$ are two integers such that the integer solutions of the system of inequalities

    $$9x - a \geq 0, \quad \text{and} \quad 8x - b < 0$$

    are $1, 2, 3$. Find the number of the ordered pairs $(a, b)$.

## Testing Questions   (B)

1.  (CHINA/2002) Given $0 \leq a - b \leq 1, 1 \leq a + b \leq 4$. Find the value of $8a + 2002b$ when the value of $a - 2b$ is maximum.

2.  Find all the positive integer-valued solutions $(x, y, z)$ of the system of inequalities

    $$\begin{cases} 3x + 2y - z & = & 4, \\ 2x - y + 2z & = & 6, \\ x + y + z & < & 7. \end{cases}$$

3.  (CHINA/2004) If $x > z, y > z$, then which is always true in the following inequalities?

    (A) $x + y > 4z$,   (B) $x + y > 3z$,   (C) $x + y > 2z$,   (D) $x + y > z$.

4.  (CHINA/2003) Given that the integer solutions of the inequality $0 \leq ax + 5 \leq 4$ for $x$ are $1, 2, 3, 4$. Find the range of the constant $a$.

5.  $a, b$ are positive integers. Find the fraction $\dfrac{a}{b}$ satisfying $\dfrac{8}{9} < \dfrac{a}{b} < \dfrac{9}{10}$, and such that $b$ is minimum.

# Lecture 28

# Quadratic Inequalities and Fractional Inequalities

In this chapter the quadratic inequalities and fractional inequalities of single variable are discussed.

Any quadratic inequality can be arranged in one of the following forms

(i) $f(x) > 0$;    (ii) $f(x) \geq 0$;    (iii) $f(x) < 0$;    (iv) $f(x) \leq 0$,

where $f(x) = ax^2 + bx + c$ with $a \neq 0$. Below for the convenience of discussion, we assume $a > 0$, i.e. the curve of the quadratic function $y = ax^2 + bx + c$ is a parabola opening upwards.

### Basic Methods for Solving Quadratic Inequalities

(I)    When the equation $f(x) = 0$ has two real roots $x_1 \leq x_2$, the solution set of the inequalities (i) to (iv) are

       (i) $x < x_1$ or $x > x_2$;      (ii) $x \leq x_1$ or $x \geq x_2$;
       (iii) $x_1 < x < x_2$;        (iv) $x_1 \leq x \leq x_2$;

respectively.

Geometrically, $x_1, x_2$ are the $x$-coordinates of the points of intersection of the curve $y = f(x)$ with the $x$-axis, and the solution set is the range of the $x$-coordinates of the points on the curve with positive $y$-coordinates (for (i)), or with non-negative $y$-coordinates (for (ii)), with negative $y$-coordinates (for (iii)), or with non-positive $y$-coordinates (for (iv)).

(II)    When the equation $f(x) = 0$ has no real solution, then $f(x) > 0$ for any real $x$, so the solution set of inequalities (i) and (ii) are both the whole real axis, and no solution for (iii) and (iv).

Geometrically, the equation $f(x) = 0$ has no real solution means the whole curve of $y = f(x)$ is above the $x$-axis, so it does not intersect with the $x$-axis, hence any point on the curve has a positive $y$-coordinate.

(III)   For fractional inequalities of the forms

(i) $g(x) > 0$;    (ii) $g(x) \geq 0$;    (iii) $g(x) < 0$;    (iv) $g(x) \leq 0$,

where $g(x) = \dfrac{x-a}{x-c}$ with $a \neq c$, the problems will become those discussed in (I) when multiplying both sides of the inequality by $(x-c)^2$.

(IV)    If there are more than one linear factors in the denominator or numerator of the fractional expression $g(x)$, then it is needed to discuss the sign of $g(x)$ by partitioning the range of $x$ into several intervals, where the partition points are given by letting each linear factor be zero.

Notice that, if the factor $(x-a)^{2n+1}$ with $n \geq 1$ occurs in the numerator or denominator of $g(x)$, it can be replaced by $(x-a)$ without changing the solution set, and if $(x-a)^{2n}$ with $n \geq 1$ occurs in the the the numerator of $g(x)$, it can be removed at first, and then determine if $a$ is in the resultant solution set. If $(x-a)^{2n}$ occurs in the denominator of $g(x)$, then remove it first, and remove the point $a$ from the resultant solution set if any.

Thus, the construction of $g(x)$ to be considered can be simplified much.

(V)     In mathematical olympiad competitions, inequalities containing parameters often occur. One kind of the common problems is to determine the values or ranges of the parameters based on information contained in the given inequality and other given conditions.

## Examples

**Example 1.** Solve the inequality $(x-2)^4(x-5)^5(x+3)^3 < 0$.

   **Solution**   Since $(x-5)^5(x+3)^3 < 0 \Leftrightarrow (x-5)(x+3) < 0$, so the solution set is

$$\{-3 < x < 5\} - \{2\}, \quad \text{or equivalently,} \quad \{-2 < x < 2\} \cup \{2 < x < 5\}.$$

**Example 2.** Solve the inequality $(x^2 - x - 1)^2 \geq (x^2 + x - 3)^2$.

   **Solution**   By factorization the given inequality can be simplified.

$$(x^2 - x - 1)^2 \geq (x^2 + x - 3)^2 \Leftrightarrow (x^2 - x - 1)^2 - (x^2 + x - 3)^2 \geq 0$$
$$\Leftrightarrow -(2x - 2)(2x^2 - 4) \geq 0 \Leftrightarrow (x-1)(x^2 - 2) \leq 0.$$

Thus, the solution set is $\{x \le -\sqrt{2}\} \cup \{1 \le x \le \sqrt{2}\}$.

**Example 3.** Solve the inequality $\dfrac{x^2 - x - 2}{x^2 - 3x + 1} \ge 0$.

**Solution** The inequality is equivalent to the systems

$$x^2 - x - 2 \ge 0, x^2 - 3x + 1 > 0 \quad \text{or} \quad x^2 - x - 2 \le 0, x^2 - 3x + 1 < 0.$$

$$x^2 - x - 2 \ge 0, \quad x^2 - 3x + 1 > 0$$
$$\Leftrightarrow (x - 2)(x + 1) \ge 0, \quad (x - \frac{3 - \sqrt{5}}{2})(x - \frac{3 + \sqrt{5}}{2}) > 0$$
$$\Leftrightarrow \{\{x \le -1\} \cup \{x \ge 2\}\} \cap \{\{x < \frac{3 - \sqrt{5}}{2}\} \cup \{x > \frac{3 + \sqrt{5}}{2}\}\}$$
$$\Leftrightarrow \{x \le -1\} \cup \{x > \frac{3 + \sqrt{5}}{2}\}.$$
$$x^2 - x - 2 \le 0, \quad x^2 - 3x + 1 < 0$$
$$\Leftrightarrow (x - 2)(x + 1) \le 0, (x - \frac{3 - \sqrt{5}}{2})(x - \frac{3 + \sqrt{5}}{2}) < 0$$
$$\Leftrightarrow \{\{-1 \le x \le 2\}\} \cap \{\{\frac{3 - \sqrt{5}}{2} < x < \frac{3 + \sqrt{5}}{2}\}\}$$
$$\Leftrightarrow \{\frac{3 - \sqrt{5}}{2} < x \le 2\}.$$

Thus, the solution set is $\{x \le -1\} \cup \{\dfrac{3 - \sqrt{5}}{2} < x \le 2\} \cup \{x > \dfrac{3 + \sqrt{5}}{2}\}$.

**Example 4.** Solve the inequality $\dfrac{x - 2}{x + 3} > -1$.

**Solution** $\dfrac{x - 2}{x + 3} > -1 \Leftrightarrow \dfrac{2x + 1}{x + 3} > 0 \Leftrightarrow (2x + 1)(x + 3) > 0$, so the solution set is

$$\{x < -3\} \cup \{x > -\frac{1}{2}\}.$$

**Example 5.** Solve the inequality $\dfrac{(x + 1)(x - 2)}{(x - 4)} > 0$, where $x \ne 4$.

**Solution** List the following table

| Range of $x$ | $x < -1$ | $-1 < x < 2$ | $2 < x < 4$ | $4 < x$ |
|---|---|---|---|---|
| Sign of $\dfrac{(x + 1)(x - 2)}{x - 4}$ | $-$ | $+$ | $-$ | $+$ |

Therefore the solution set is $S = (-1, 2) \cup (4, +\infty)$.

**Example 6.** Find the solution set of the inequality $\dfrac{x+1}{x-1} > \dfrac{6}{x}$. ($x \neq 1$ and $x \neq 0$.)

**Solution**   Change $\dfrac{x+1}{x-1} > \dfrac{6}{x}$ to the form $\dfrac{x+1}{x-1} - \dfrac{6}{x} > 0$, then

$$\dfrac{x+1}{x-1} - \dfrac{6}{x} > 0 \;\Leftrightarrow\; \dfrac{x(x+1) - 6(x-1)}{x(x-1)} > 0 \Leftrightarrow \dfrac{x^2 - 5x + 6}{x(x-1)} > 0,$$

$$\Leftrightarrow \dfrac{(x-2)(x-3)}{x(x-1)} > 0.$$

Based on the following table

| Range of $x$ | $(-\infty, 0)$ | $(0, 1)$ | $(1, 2)$ | $(2, 3)$ | $(3, +\infty)$ |
|---|---|---|---|---|---|
| Sign of $\dfrac{(x-2)(x-3)}{x(x-1)}$ | $+$ | $-$ | $+$ | $-$ | $+$ |

the solution set is

$$(-\infty, 0) \cup (1, 2) \cup (3, +\infty).$$

**Example 7.** Solve the quadratic inequality $ax^2 - (a+1)x + 1 < 0$, where $a$ is a parameter.

**Solution**   Since $a \neq 0$, $ax^2 - (a+1)x + 1 < 0 \Leftrightarrow a(x - \dfrac{1}{a})(x-1) < 0$.

(i)     If $a > 1$, the solution set is $\dfrac{1}{a} < x < 1$.

(ii)    If $a = 1$, no solution.

(iii)   If $0 < a < 1$, the solution set is $1 < x < \dfrac{1}{a}$.

(iv)    If $a < 0$, the inequality becomes $(x - \dfrac{1}{a})(x-1) > 0$, the solution set is

$$\{x < \dfrac{1}{a}\} \cup \{x > 1\}.$$

**Example 8.** Given that the inequality $kx^2 - kx - 1 < 0$ holds for any real $x$, then
(A) $-4 < k \leq 0$,  (B) $-4 \leq k \leq 0$,  (C) $-4 < k < 0$,  (D) $-4 \leq k < 0$.

**Solution**   It is obvious that the inequality holds when $k = 0$.
When $k < 0$, the curve of $y = kx^2 - kx - 1$ is open downwards, and is below the $x$ axis, so the equation $kx^2 - kx - 1 = 0$ has no real roots, i.e. its discriminant is negative.

$$\Delta = k^2 + 4k < 0 \Leftrightarrow k > -4.$$

Thus, $-4 < k \le 0$, the answer is (A).

**Example 9.** Given that the solution set of the quadratic inequality $ax^2 + bx + c > 0$ is $1 < x < 2$. Find the solution set of the inequality $cx^2 + bx + a < 0$.

**Solution** The first inequality has the solution set $1 < x < 2$ implies that $a < 0$, and

$$x^2 + \frac{b}{a}x + \frac{c}{a} < 0 \Leftrightarrow (x - 1)(x - 2) < 0 \Leftrightarrow x^2 - 3x + 2 < 0.$$

Therefore $\frac{b}{a} = -3$, $\frac{c}{a} = 2$ or $b = -3a, c = 2a$ and $a < 0$. Then the second inequality becomes

$$a(2x^2 - 3x + 1) < 0, \quad \text{i.e. } 2x^2 - 3x + 1 > 0.$$

Thus, $(2x - 1)(x - 1) > 0$, and its solution set is $\{x < \frac{1}{2}\} \cup \{x > 1\}$.

## Testing Questions    (A)

1.  Solve the inequality $(2 + x)(x - 5)(x + 1) > 0$.

2.  Solve the inequality $x^2(x^2 - 4) < 0$.

3.  Solve the inequality $x^3 \le 6x - x^2$.

4.  Solve the inequality $x - 1 > (x - 1)(x + 2)$.

5.  Solve the inequality $(x^3 - 1)(x^3 + 1) > 0$.

6.  Solve the inequality $\dfrac{2x - 4}{x + 3} > \dfrac{x + 2}{2x + 6}$.

7.  Find the solution set of the inequality $\dfrac{x}{x + 2} \ge \dfrac{1}{x}$.

8.  Find the solution set of the inequality $\dfrac{x - 1}{x^2} \le 0$.

9.  Find the solution set of the inequality $\dfrac{x(2x - 1)^2}{(x + 1)^3(x - 2)} > 0$.

10.  Find the solution set of the inequality $\dfrac{2x^2}{x + 1} \ge x$.

## Testing Questions    (B)

1.  (CHINA/2002) Find the positive solutions of the inequality

$$\frac{x^2+3}{x^2+1} + \frac{x^2-5}{x^2-3} \geq \frac{x^2+5}{x^2+3} + \frac{x^2-3}{x^2-1}.$$

2.  (CHINA/2001) Solve the fractional inequality

$$\frac{x+2}{4x+3} - \frac{x}{4x+1} > \frac{x}{4x-1} - \frac{x-2}{4x-3}.$$

3.  Given that $a, b$ are positive constant with $a < b$. If the inequality

$$a^3 + b^3 - x^3 \leq (a+b-x)^3 + m$$

    holds for any real $x$, find the minimum value of the parameter $m$.

4.  Given that the quadratic function $f(x) = x^2 - 2ax + 6 \geq a$ for $-2 \leq x \leq 2$, find the range of the constant $a$.

5.  Given that the inequality

$$\frac{1}{8}(2a - a^2) \leq x^2 - 3x + 2 \leq 3 - a^2$$

    holds for any real $x$ in the interval $[0, 2]$. Find the range of the parameter $a$.

# Lecture 29

# Inequalities with Absolute Values

The most important technique for solving inequalities with absolute values is the same as in solving equations with absolute values, i.e., it is necessary to remove the absolute value signs, such that they become normal inequalities to solve.

### Basic Methods for Removing Absolute Value Signs

(I)     $|a| \leq b$ is equivalent to $-b \leq a \leq b$.
        Notice that, here it is not needed to let $b$ be non-negative, since if $b < 0$,
        then $-b \leq a \leq b$ has no solution for $a$.

(II)    $|a| \geq |b|$ is equivalent to $a^2 \geq b^2$.

(III)   $|a| \geq b$ is equivalent to $a \leq -b$ or $a \geq b$.

(IV)    For simplifying the given inequality and removing absolute value signs, some substitutions of variables or expressions, like $y = |x|$, is useful.

(V)     When two or more pairs of absolute value signs occur in a *same layer* of the given inequality, the general method for removing the absolute value signs is partitioning the range of variable into several intervals, so that the values of each expression between a pair of absolute signs has a fixed sign. For this it is necessary to **let each such expression be zero, and take its roots as the partition points**. For example, for solving the inequality

$$|x - 2| + |x - 4| < 3,$$

then the number axis is partitioned into three intervals $(-\infty, 2]$, $(2, 4]$ and $(4, \infty)$, where 2 and 4 are obtained by letting $x - 2 = 0$ and $x - 4 = 0$.

### Examples

**Example 1.** Solve the inequality $|x^2 - 2x + 5| < 4$.

89

**Solution**

$$|x^2 - 2x + 5| < 4 \Leftrightarrow -4 < x^2 - 6x + 5 < 4$$
$$\Leftrightarrow 0 < x^2 - 6x + 9 \text{ and } x^2 - 6x + 1 < 0.$$

$0 < x^2 - 6x + 9 \Leftrightarrow 0 < (x-3)^2 \Leftrightarrow x \neq 3$, and

$$x^2 - 6x + 1 < 0 \Leftrightarrow 3 - 2\sqrt{2} < x < 3 + 2\sqrt{2},$$

so the solution set is $\{3 - 2\sqrt{2} < x < 3 + 2\sqrt{2}\} - \{3\}$.

**Example 2.** Solve the inequality $|3x^2 - 2| < 1 - 4x$

**Solution**

$$|3x^2 - 2| < 1 - 4x \Leftrightarrow 4x - 1 < 3x^2 - 2 < 1 - 4x$$
$$\Leftrightarrow 0 < 3x^2 - 4x - 1 \text{ and } 3x^2 + 4x - 3 < 0$$
$$\Leftrightarrow x < \frac{2 - \sqrt{7}}{3} \text{ or } x > \frac{2 + \sqrt{7}}{3} \text{ and } \frac{-2 - \sqrt{13}}{3} < x < \frac{-2 + \sqrt{13}}{3}$$

Since $\dfrac{-2 - \sqrt{13}}{3} < \dfrac{2 - \sqrt{7}}{3} < \dfrac{-2 + \sqrt{13}}{3} < \dfrac{2 + \sqrt{7}}{3}$, the solution set is

$$\frac{-2 - \sqrt{13}}{3} < x < \frac{2 - \sqrt{7}}{3}.$$

**Example 3.** Solve the inequality $\left|\dfrac{x+1}{x-1}\right| \leq 1$.

**Solution**   The given inequality implies that $x \neq 1$ and $|x - 1| > 0$, so

$$\left|\frac{x+1}{x-1}\right| \leq 1 \Leftrightarrow |x + 1| \leq |x - 1| \Leftrightarrow (x + 1)^2 \leq (x - 1)^2$$
$$\Leftrightarrow x^2 + 2x + 1 \leq x^2 - 2x + 1 \Leftrightarrow 4x \leq 0 \Leftrightarrow x \leq 0.$$

Thus, the solution set is $\{x \leq 0\}$.

**Example 4.** Solve the inequality $|x^2 - x| > 2$.

**Solution**   Since $|x^2 - x| > 2 \Leftrightarrow x^2 - x < -2$ or $x^2 - x > 2$, and

$$x^2 - x < -2 \Leftrightarrow x^2 - x + 2 < 0, \quad x^2 - x > 2 \Leftrightarrow x^2 - x - 2 > 0.$$

The inequality $x^2 - x + 2 < 0$ has no real solution since the discriminant $\Delta$ of the corresponding equation $x^2 - x + 2 = 0$ is negative:

$$\Delta = (-1)^2 - 8 = -7 < 0,$$

so the curve $y = x^2 - x + 2$ is above the $x$ axis entirely. However

$$x^2 - x - 2 > 0 \Leftrightarrow (x - 2)(x + 1) > 0 \Leftrightarrow x < -1 \text{ or } x > 2.$$

Thus the solution set is $(-\infty, -1) \cup (2, +\infty)$.

**Example 5.** Solve the inequality $x^2 - 2x - 5|x - 1| + 7 \le 0$.

**Solution** Since

$$x^2 - 2x - 5|x - 1| + 7 = (x^2 - 2x + 1) - 5|x - 1| + 6 = (x - 1)^2 - 5|x - 1| + 6,$$

let $y = |x - 1|$, the given inequality then becomes

$$y^2 - 5y + 6 \le 0,$$
$$(y - 2)(y - 3) \le 0,$$
$$\therefore 2 \le y \le 3.$$

Returning to $x$, it becomes $2 \le |x - 1| \le 3$.

By solving $2 \le |x - 1|$, then $x - 1 \le -2$ or $x - 1 \ge 2$, so its solution set is $\{x \le -1\} \cup \{x \ge 3\}$.

By solving $|x - 1| \le 3$, then $-3 \le x - 1 \le 3$, so its solution set is $-2 \le x \le 4$.

By taking the common part of these two solution sets, the solution set of the original inequality is

$$\{-2 \le x \le -1\} \cup \{3 \le x \le 4\}.$$

**Example 6.** Solve the inequality $\dfrac{3x^2 - 8|x| - 3}{x^2 + 2x + 3} > 0$.

**Solution** Since $x^2 + 2x + 3 = (x + 1)^2 + 2 \ge 2 > 0$ for any real $x$, the solution set of the given inequality is equivalent to that of the inequality $3x^2 - 8|x| - 3 > 0$. For solving it, let $y = |x|$, then it follows that

$$3y^2 - 8y - 3 > 0,$$
$$(3y + 1)(y - 3) > 0,$$
$$\therefore y > 3 \text{ or } 3y + 1 < 0 \text{ (not acceptable since } y \ge 0).$$

Returning to $x$, the solution set is $\{|x| > 3\}$ or equivalently, $\{x < -3\} \cup \{x > 3\}$.

**Example 7.** Solve the inequality $2|x| - |x - 2| \ge 0$.

**Solution**　Since $2|x| - |x - 2| \geq 0 \Leftrightarrow 2|x| \geq |x - 2|$, by taking squares to both sides, it follows that

$$4x^2 \geq (x - 2)^2,$$
$$4x^2 \geq x^2 - 4x + 4,$$
$$3x^2 + 4x - 4 \geq 0,$$
$$(3x - 2)(x + 2) \geq 0.$$

Since the roots of the quadratic equation $(3x - 2)(x + 2) = 0$ are $\dfrac{2}{3}$ and $-2$, the solution set of the quadratic inequality is $\{x \leq -2\} \cup \{x \geq \dfrac{2}{3}\}$.

**Note:**　This problem can be solved by partitioning the number axis into three parts: $x \leq 0, 0 < x \leq 2$, and $2 < x$.

**Example 8.** Solve the inequality $|2x - 1| - |x + 1| > 2$.

**Solution**　It's needed to partition the real axis into three intervals by using the partition points $-1$ and $\dfrac{1}{2}$.

For $x \leq -1$, the inequality becomes $(1 - 2x) + (x + 1) > 2$ i.e. $x < 0$, so $(-\infty, -1]$ is in the solution set.

For $-1 < x \leq \dfrac{1}{2}$, the inequality becomes $(1 - 2x) - (x + 1) > 2$, the solution set is $-1 < x < -\dfrac{2}{3}$.

For $x > \dfrac{1}{2}$, the inequality becomes $(2x - 1) - (x + 1) > 2$, so the solution set is $x > 4$.

Thus, the solution set is $(-\infty, -\dfrac{2}{3}) \cup (-4, +\infty)$.

**Example 9.** (CHNMOL/1995) Given that the real numbers $a, b$ satisfy the inequality

$$||a| - (a + b)| < |a - |a + b||,$$

then

(A) $a > 0, b > 0$;　(B) $a < 0, b > 0$;　(C) $a > 0, b < 0$;　(D) $a < 0, b < 0$.

**Solution**　There are two layers of absolute value signs in the given expression. By taking squares to both sides, the absolute value signs in the outer layer can be removed.

$$||a| - (a + b)| < |a - |a + b|| \Leftrightarrow (|a| - (a + b))^2 < (a - |a + b|)^2,$$
$$a^2 + (a + b)^2 - 2|a|(a + b) < a^2 + (a + b)^2 - 2a|a + b|,$$
$$a|a + b| < |a|(a + b),$$

therefore $a, a + b$ are both not zero, so $\dfrac{a}{|a|} < \dfrac{a+b}{|a+b|}$. Since both sides have absolute value 1, so $a < 0$ and $a + b > 0$, thus, $a < 0, b > 0$, the answer is (B).

When an inequality with absolute values contains parameters, then, similar to the case of without absolute values, the inequality can give some information on the range of the parameters, as shown in the following example.

**Example 10.** (AHSME/1964) When $x$ is a real number and $|x - 4| + |x - 3| < a$, where $a > 0$, then

(A) $0 < a < 0.01$, (B) $0.01 < a < 1$, (C) $0 < a < 1$,

(D) $0 < a \leq 1$, (E) $1 < a$.

**Solution** Use 3 and 4 as the partition points to partition the real axis as three intervals,

 (i) for $x \leq 3$, then $|x - 4| + |x - 3| = 4 - x + 3 - x = 7 - 2x \geq 1$,
 (ii) for $3 < x \leq 4$, then $|x - 4| + |x - 3| = 4 - x + x - 3 = 1$,
 (iii) for $4 < x$, then $|x - 4| + |x - 3| = x - 4 + x - 3 = 2x - 7 > 1$,
so the left hand side must be at least 1. Thus, $a > 1$, the answer is (E).

## Testing Questions   (A)

1.   Find the solution set of the inequality $|x^2 + x + 1| \leq 1$.

2.   Find the solution set of the inequality $|3 - 2x| \leq |x + 4|$.

3.   Find the solution set of the inequality $\left| \dfrac{x + 1}{x - 1} \right| \geq 1$.

4.   Find the solution set of the inequality $|x + 3| > 2x + 3$.

5.   Solve the inequality $|x^2 - 4x - 5| > x^2 - 4x - 4$.

6.   Solve the inequality $|x + 1| > \dfrac{2}{x}$.

7.   Solve the inequality $|x + 1| + |x - 2| \leq 3x$.

8.   Solve the inequalities   (i) $\dfrac{6}{|x| + 1} < |x|$;   (ii) $\dfrac{1 - |x|}{3|x| - 6} > 0$.

9.   Solve the inequality $\dfrac{|x|}{x} < |x^2 - 1| \ (x \neq 0)$.

10.  (CHINA/2005) Given that the rational numbers $a, b$ satisfy the inequality

$$||a| + (a - b)| > |a + |a - b||.$$

Determine which of the following listed expressions holds.

(A) $a > 0, b > 0$;   (B) $a > 0, b < 0$;   (C) $a < 0, b > 0$;   (D) $a < 0, b < 0$.

## Testing Questions    (B)

1.  (CHINA/2003) Given that the real numbers $a, b, c$ satisfy $a + b + c = 2, abc = 4$.

(i) Find the minimum value of the maximal value of $a, b, c$

(ii) Find the minimum value of $|a| + |b| + |c|$.

2.  (CHINA/2001) When $|a| < |c|, b = \dfrac{a + c}{2}, |b| < 2|a|, S_1 = \left|\dfrac{a - b}{c}\right|, S_2 = \left|\dfrac{b - c}{a}\right|, S_3 = \left|\dfrac{a - c}{b}\right|$, then the relation of the sizes of $S_1, S_2, S_3$ is

(A) $S_1 < S_2 < S_3$,       (B) $S_1 > S_2 > S_3$,

(C) $S_1 < S_3 < S_2$,       (D) $S_1 > S_3 > S_2$.

3.  (CHINA/2004) Solve inequality $\dfrac{|2y + |y| + 10|}{|6y + 2|y| + 5|} > 1$

4.  Given (i) $a > 0$; (ii) $|ax^2 + bx + c| \le 1$ if $-1 \le x \le 1$; (iii) $ax + b$ has its maximum value 2 when $-1 \le x \le 1$. Find the values of constants $a, b, c$.

5.  Given that the real numbers $a \le b \le c$ satisfy $ab + bc + ca = 0, abc = 1$. Find the maximum real number $k$, such that

$$|a + b| \ge k|c|$$

holds for any such $a, b, c$.

# Geometric Inequalities

In this chapter we discuss the inequalities that deal with lengths of segments, sizes of angles and sizes of areas of geometric graphs. The following theorems are the basic tools for dealing with geometric inequalities:

**Theorem I.** *Among the paths joining two given points, the segment joining them is the shortest.*

**Theorem II.** *For a straight line $\ell$ and a point $P$ outside $\ell$, if $Q$ is the foot of the perpendicular from $P$ to $\ell$, and $A, B$ are other two points on $\ell$, such that $AQ < BQ$, then $PQ < PA < PB$.*

**Theorem III (Triangle Inequality).** *For any triangle $ABC$, let $BC = a, CA = b, AB = c$, then $a < b+c, b < c+a, c < a+b$, or equivalently, $a > |b-c|, b > |c - a|, c > |a - b|$.*

**Theorem IV.** *For a triangle, a longer side is opposite to a bigger angle, a shorter side is opposite to a smaller angle, and vice versa.*

**Theorem V.** *For any triangle, the median to a side is less than half of the sum of other two sides.*

## Examples

**Example 1.** Given that the point $R$ is an inner point of $\triangle ABC$, prove that $AB + AC > BR + RC$.

**Solution**  Extending $BR$ to intersect $AC$ at $S$.
Then

$$
\begin{aligned}
AB + AC &= AB + AS + SC \\
&> BS + SC = BR + (RS + SC) \\
&> BR + RC.
\end{aligned}
$$

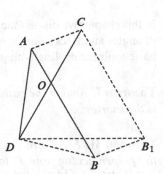

**Example 2.** (RUSMO/1993) Let $AB$ and $CD$ be two segments of length 1. If they intersect at $O$ such that $\angle AOC = 60°$, prove that $AC + BD \geq 1$.

**Solution**  Connect $AC, BD$, and introduce $CB_1$
$\parallel AB$, where $CB_1 = AB$. Then $ABB_1C$ is a parallelogram, so $BB_1 = AC$. Connect $B_1D$. Then $\triangle CDB_1$ is equilateral. Applying the triangle inequality to $\triangle DBB_1$ gives

$$
AC + BD = BB_1 + BD > DB_1 = CD = 1,
$$

as desired.

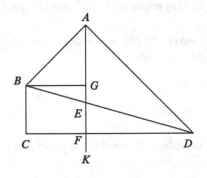

Note that when $A$ and $C$ coincide, then $AC + BD = BD = 1$.

**Example 3.** (CHINA/1994)  In the convex quadrilateral $ABCD$, $\angle BAD = \angle BCD = 90°$, and the ray $AK$ bisects the $\angle BAD$. If $AK \parallel BC, AK \perp CD$, and $AK$ intersects $BD$ at $E$, prove that $AE < \frac{1}{2}CD$.

**Solution**  Suppose that $AK$ intersects
$CD$ at $F$ and $BG \perp AK$ at $G$. Then
$\triangle ABG$ and $\triangle ADF$ are both isosceles
right triangles, so

$$
BG = AG, FD = AF.
$$

Since $BCFG$ is a rectangle,
$FD = AF > AG = BG = CF$, so

$$
FD > \frac{1}{2}(CF + FD) = \frac{1}{2}CD,
$$

$$
\therefore EF = \frac{FD}{CD} \cdot BC > \frac{1}{2}BC.
$$

Since $BC + CF = FD$, we have

$$CD = CF + FD = 2FD - BC > 2(FD - EF) = 2(AF - EF) = 2AE,$$

i.e. $AE < \dfrac{1}{2}CD$.

**Example 4.** (CHINA/1990) For $\triangle ABC$, let $a = BC, b = CA, c = AB$. If $b < \dfrac{1}{2}(a + c)$, prove that $\angle B < \dfrac{1}{2}(\angle A + \angle C)$.

**Solution** Considering $\angle A + \angle B + \angle C = 180°$,

$$\angle B < \dfrac{1}{2}(\angle A + \angle C) \iff \angle B < 60°.$$

Now construct an isosceles triangle $BDE$ by extending $BA$ to $D$ and extending $BC$ to $E$, such that $BD = c + a, BE = a + c$. Next, we make parallelogram $ACFD$ such that $AC \parallel DF, AD \parallel CF$, then

$$\triangle BAC \cong \triangle CFE, \therefore FE = AC = FD = b.$$

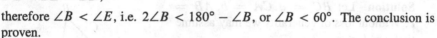

From $DE < DF + FE = 2b < a + c$, it follows that $DE < BE = BD$,

therefore $\angle B < \angle E$, i.e. $2\angle B < 180° - \angle B$, or $\angle B < 60°$. The conclusion is proven.

**Example 5.** (KIEV/1967) Given that the longest side $AC$ of $\triangle ABC$ satisfies $AC > BC$. If the point $M$ is on the extension of of $AC$ such that $CM = BC$. Prove that $\angle ABM > 90°$.

**Solution** $AC > BC$ implies $\angle ABC > \angle A$. Connect $BM$. Then

$$\angle ABM = \angle ABC + \angle CBM$$

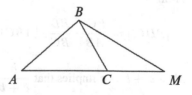

$$= \dfrac{1}{2}\angle ABC + \dfrac{1}{2}\angle ABC + \dfrac{1}{2}\angle ACB$$

$$= \dfrac{1}{2}\angle ABC - \dfrac{1}{2}\angle A + 90° > 90°.$$

Thus, $\angle ABM$ is obtuse.

**Example 6.** (MOSCOW/1951)  Given that two convex quadrilaterals $ABCD$ and $A'B'C'D'$ have equal corresponding sides, i.e. $AB = A'B', BC = B'C'$, etc. Prove that if $\angle A > \angle A'$, then $\angle B < \angle B', \angle C > \angle C', \angle D < \angle D'$.

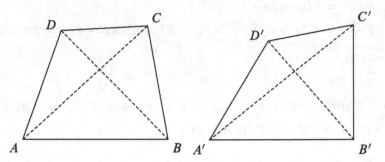

**Solution** In $\triangle ABD$ and $\triangle A'B'D'$, $AB = A'B', AD = A'D', \angle A > \angle A'$ yields $BD > B'D'$. In $\triangle BCD$ and $\triangle B'C'D'$, $BC = B'C', CD = C'D'$ and $BD > B'D'$ implies $\angle C > \angle C'$.

Therefore $\angle B + \angle D < \angle B' + \angle D'$. Suppose $\angle B \geq \angle B'$, then $\angle D < \angle D'$, so by similar reasoning, $\angle B' > \angle B$, a contradiction. Thus, $\angle B < \angle B'$ and $\angle D < \angle D'$, as desired.

**Example 7.** (USAMO/1996) Let $ABC$ be a triangle. Prove that there is a line $\ell$ (in the plane of triangle $ABC$) such that the intersection of the interior of triangle $ABC$ and the interior of its reflection $A'B'C'$ in $\ell$ has area more than $2/3$ the area of triangle $ABC$.

**Solution** Let $BC = a, CA = b, AB = c$. Without loss of generality we may assume that $a \leq b \leq c$.

Let $AD$ be the angle bisector of the $\angle BAC$, $B', C'$ be the symmetric points of $B, C$ in the line $AD$, respectively, then $C'$ is on the segment $AB$ and $B'$ is on the extension of $AC$, as shown in the figure.

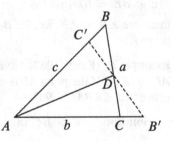

From

$$[BDC'] = \frac{BC'}{AB} \cdot \frac{BD}{BC} \cdot [ABC] = \frac{c-b}{c} \cdot \frac{c}{b+c}[ABC] = \frac{c-b}{b+c}[ABC].$$

$2b > a + b > c$ implies that $\dfrac{b}{c+b} > \dfrac{b}{2b+b} = \dfrac{1}{3}$, so

$$[AC'DC] = \left(1 - \frac{c-b}{c+b}\right)[ABC] = \frac{2b}{c+b}[ABC] > \frac{2}{3}[ABC].$$

Thus, the line $AD$ satisfies the requirement.

**Example 8.** (CHINA/1978) Through the center of gravity $G$ of the $\triangle ABC$ introduce a line to divide $\triangle ABC$ into two parts. Prove that the difference of areas of the two parts is not greater than $\frac{1}{9}$ of area of the $\triangle ABC$.

**Solution** Suppose that an arbitrary line passing through $G$ intersects $AB$ and $AC$ at $D$ and $E$ respectively.

When the line $DE$ is parallel to $BC$, then

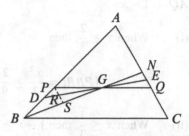

$$[ADE] = \left(\frac{DE}{BC}\right)^2 [ABC] = \frac{4}{9}[ABC], \text{ so}$$

$$|[ADE] - [DBCE]| = \left(\frac{5}{9} - \frac{4}{9}\right)[ABC]$$

$$= \frac{1}{9}[ABC], \text{ the conclusion is true.}$$

When $DE$ is not parallel to $BC$, say $D$ is between $P$ and $B$ and $E$ is between $Q$ and $N$, where $PQ \parallel BC$ and $N$ is the midpoint of $AC$, as shown in the figure, we show that

$$|[DBCE] - [ADE]| < |[PBCQ] - [APQ]| = \frac{1}{9}[ABC]$$

below. Since $\angle DPG > \angle PQA$, we can introduce $PS \parallel AC$, intersecting $DE$ and $BN$ at $R$ and $S$ respectively. Then $\triangle PRG \cong \triangle QEG$ and $\triangle RSG \cong \triangle ENG$, so that

$$[PRG] = [QEG] \quad \text{and} \quad [RSG] = [ENG].$$

Therefore $[DBCE] = [PBCQ] - [PDG] + [QEG] < [PBCQ]$ and $[ADE] = [APQ] + [PDG] - [QEG] > [APQ]$, so that

$$[DBCE] - [ADE] < [PBCQ] - [APQ] = \frac{1}{9}[ABC].$$

It suffices to show that $[DBCE] > [ADE]$. For this notice that

$$\begin{aligned} [DBCE] &= [BCN] - [ENG] + [DBG] > [BCN] - [ENG] + [RSG] \\ &= [BCN] = \frac{1}{2}[ABC], \end{aligned}$$

$$\therefore [ADE] < \tfrac{1}{2}[ABC] < [DBCE].$$

Thus, the conclusion is proven.

**Example 9.** (CMO/1969) Let $\triangle ABC$ be the right-angled isosceles triangle whose equal sides have length 1. $P$ is a point on the hypotenuse, and the feet of the perpendiculars from $P$ to the other sides are $Q$ and $R$. Consider the areas of the triangles $APQ$ and $PBR$, and the area of the rectangle $QCRP$. Prove that regardless of how $P$ is chosen, the largest one of these three areas is at least $\dfrac{2}{9}$.

    **Solution**  Let $BR = x$, then $BR = PR = QC = x$ and $RC = PQ = AQ = 1 - x$.

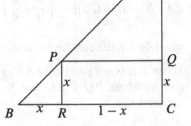

(i)    When $x \geq \dfrac{2}{3}$, then

$$[PBR] = \frac{x^2}{2} \geq \frac{2}{9}.$$

(ii)    When $x \leq \dfrac{1}{3}$, then $1 - x \geq \dfrac{2}{3}$, so that

$$[APQ] = \frac{(1-x)^2}{2} \geq \frac{2}{9}.$$

(iii)    When $\dfrac{1}{3} < x < \dfrac{2}{3}$, then $-\dfrac{1}{6} < x - \dfrac{1}{2} < \dfrac{1}{6}$, so that

$$[QCRP] = x(1-x) = -x^2 + x = -\left(x - \frac{1}{2}\right)^2 + \frac{1}{4} > -\frac{1}{36} + \frac{1}{4} = \frac{2}{9}.$$

Thus, the conclusion is proven.

# Testing Questions   (A)

1.    Given that $R$ is an inner point of $\triangle ABC$. Prove that

$$\frac{1}{2}(AB + BC + CA) < RA + RB + RC < AB + BC + CA.$$

2.    (BMO/1967) In $\triangle ABC$, if $\angle C > \angle B$, and $BE, CF$ are the heights on $CA$ and $AB$ respectively. Prove that $AB + CF > AC + BE$.

3.    In $\triangle ABC$, $AB > BC$, $AD \perp BC$ at $D$. $P$ is an arbitrary point on $AD$ different from $A$ and $D$, prove that $PB - PC > AB - AC$.

4.  For an acute triangle $ABC$, let $a = BC, b = CA, c = AB$, and the lengths of height on $BC, CA, AB$ are $h_a, h_b, h_c$ respectively, prove that

$$\frac{1}{2}(a + b + c) < h_a + h_b + h_c < a + b + c.$$

5.  (MOSCOW/1974) Prove that if three segments with lengths $a, b, c$ can form a triangle, then the three segments of lengths $\dfrac{1}{a+b}, \dfrac{1}{b+c}, \dfrac{1}{c+a}$ can form a triangle also.

6.  Given that $BB_1$ and $CC_1$ are two medians of $\triangle ABC$. Prove that $BB_1^2 + CC_1^2 > \dfrac{9}{8}BC^2$.

7.  (MOSCOW/1972) A straight line intersects the sides $AB$ and $BC$ of $\triangle ABC$ at the point $M$ and $K$ respectively, such that the area of $\triangle MBK$ and the area of the quadrilateral $AMKC$ are equal. Prove that $\dfrac{MB + BK}{AM + CA + KC} \geq \dfrac{1}{3}$.

8.  (CHINA/1994) There are $n$ straight lines in a plane, such that every two intersect with each other. Prove that among the angles formed there is at least one angle which is not greater than $\dfrac{180°}{n}$.

9.  (RUSMO/1983) In $\triangle ABC$, $D$ is the midpoint of $AB$, $E$ and $F$ are on $AC$ and $BC$ respectively. Prove that the area of $\triangle DEF$ is not greater than sum of areas of $\triangle ADE$ and $\triangle BDF$.

10. (CHNMOL/1979) Given that $\triangle ABC$ is an acute triangle, $a = BC, b = CA, c = AB$ and $a > b > c$. Among its 3 inscribed squares, which one has the maximum area?

# Testing Questions    (B)

1.  (KIEV/1969) Is there a triangle with the three altitudes of lengths $1, \sqrt{5}, 1 + \sqrt{5}$?

2.  (RUSMO/1981) The points $C_1, A_1, B_1$ belong to sides $AB, BC, CA$, respectively, of the $\triangle ABC$.

$$\frac{AC_1}{C_1B} = \frac{BA_1}{A_1C} = \frac{CB_1}{B_1A} = \frac{1}{3}.$$

Prove that the perimeter $P$ of the $\triangle ABC$ and the perimeter $p$ of the $\triangle A_1 B_1 C_1$ satisfy inequality $\dfrac{P}{2} < p < \dfrac{3}{4} P$.

3.  (KIEV/1966) Let $a, b, c$ be the lengths of three sides of $\triangle ABC$, and $I = a + b + c$, $S = ab + bc + ca$. Prove that $3S \leq I^2 \leq 4S$.

4.  (RUSMO/1989) If $a, b, c$ denote the lengths of sides of a triangle, satisfying $a + b + c = 1$, prove that

$$a^2 + b^2 + c^2 + 4abc < \frac{1}{2}.$$

5.  (PUTNAM/1973) (i) Given that $\triangle ABC$ is an arbitrary triangle, the points $X, Y, Z$ are on the sides $BC, CA, AB$ respectively. If $BX \leq XC, CY \leq YA, AZ \leq ZB$, prove that $[XYZ] \geq \frac{1}{4}[ABC]$.

    (ii) Given that $\triangle ABC$ is an arbitrary triangle, the points $X, Y, Z$ are on the sides $BC, CA, AB$ respectively (but there is no any assumptions to the ratio of distances, like $\dfrac{BX}{XC}$). Please use the method used in (i) or otherwise to show that among $\triangle AZY, \triangle BXZ, \triangle CYZ$ there must be one with area not greater than that of $\triangle XYZ$.

6.  (IREMO/2003) Let $T$ be a triangle of perimeter 2, and $a, b, c$ be the lengths of its three sides. Prove that

    (i)   $abc + \dfrac{28}{27} \geq ab + bc + ca$;

    (ii)  $ab + bc + ca \geq abc + 1$.

# Solutions to Testing Questions

# Solutions to Testing Questions

## Solutions to Testing Questions   16

### Testing Questions   (16-A)

1.  $\because x < 2, |\sqrt{(x-2)^2} + \sqrt{(3-x)^2}| = |(2-x) + (3-x)| = 5 - 2x$, so the answer is (A).

2.

$$\frac{1+\sqrt{2}+\sqrt{3}}{1-\sqrt{2}+\sqrt{3}} = \frac{(\sqrt{3}+1+\sqrt{2})^2}{[(\sqrt{3}+1)-\sqrt{2}][(\sqrt{3}+1)+\sqrt{2}]}$$

$$= \frac{(\sqrt{3}+1+\sqrt{2})^2}{(\sqrt{3}+1)^2-(\sqrt{2})^2} = \frac{6+2\sqrt{3}+2\sqrt{6}+2\sqrt{2}}{4+2\sqrt{3}-2}$$

$$= \frac{3+\sqrt{3}+\sqrt{2}+\sqrt{6}}{\sqrt{3}+1} = \frac{(\sqrt{3}+1)(\sqrt{3}+\sqrt{2})}{\sqrt{3}+1} = \sqrt{3}+\sqrt{2}.$$

3.  $\dfrac{x^2-4x+3+(x+1)\sqrt{x^2-9}}{x^2+4x+3+(x-1)\sqrt{x^2-9}} = \dfrac{(x-3)(x-1)+(x+1)\sqrt{(x-3)(x+3)}}{(x+3)(x+1)+(x-1)\sqrt{(x+3)(x-3)}}$

$$= \frac{\sqrt{x-3}[(x-1)\sqrt{x-3}+(x+1)\sqrt{x+3}]}{\sqrt{x+3}[(x+1)\sqrt{x+3}+(x-1)\sqrt{x-3}]} = \frac{\sqrt{x-3}}{\sqrt{x+3}} = \frac{\sqrt{x^2-9}}{x+3}.$$

4.  $\dfrac{2+3\sqrt{3}+\sqrt{5}}{(2+\sqrt{3})(2\sqrt{3}+\sqrt{5})} = \dfrac{(2+\sqrt{3})+(2\sqrt{3}+\sqrt{5})}{(2+\sqrt{3})(2\sqrt{3}+\sqrt{5})} = \dfrac{1}{2\sqrt{3}+\sqrt{5}}+\dfrac{1}{2+\sqrt{3}}$

$$= \frac{2\sqrt{3}-\sqrt{5}}{12-5} + \frac{2-\sqrt{3}}{4-3} = \frac{2}{7}\sqrt{3} - \frac{1}{7}\sqrt{5} + 2 - \sqrt{3} = 2 - \frac{5}{7}\sqrt{3} - \frac{1}{7}\sqrt{5}.$$

5.  $(\sqrt{5}+\sqrt{6}+\sqrt{7})(\sqrt{5}+\sqrt{6}-\sqrt{7})(\sqrt{5}-\sqrt{6}+\sqrt{7})(-\sqrt{5}+\sqrt{6}+\sqrt{7})$

$$= [(\sqrt{5}+\sqrt{6})+\sqrt{7}][(\sqrt{5}+\sqrt{6})-\sqrt{7}][\sqrt{7}-(\sqrt{6}-\sqrt{5})][\sqrt{7}+(\sqrt{6}-\sqrt{5})]$$

$$= [(\sqrt{5} + \sqrt{6})^2 - 7][7 - (\sqrt{6} - \sqrt{5})^2] = [4 + 2\sqrt{30}][-4 + 2\sqrt{30}]$$
$$= 120 - 16 = 104.$$

6. $\because a = \dfrac{(\sqrt{6})^2 - 2^2}{\sqrt{6} + 2} = \dfrac{2}{\sqrt{6} + 2}, b = \dfrac{(2\sqrt{2})^2 - (\sqrt{6})^2}{2\sqrt{2} + \sqrt{6}} = \dfrac{2}{\sqrt{8} + \sqrt{6}},$

   from $\sqrt{8} + \sqrt{6} > \sqrt{6} + 2$, we have $\therefore a > b$.

7. $a = \dfrac{1}{\sqrt{27} + \sqrt{26}}, b = \dfrac{1}{\sqrt{28} + \sqrt{27}}, c = \dfrac{1}{\sqrt{29} + \sqrt{28}}, \quad \therefore c < b < a.$

8. We have $\dfrac{3}{1 + \sqrt{3}} < x < \dfrac{3}{\sqrt{5} - \sqrt{3}} \Leftrightarrow \dfrac{3(\sqrt{3} - 1)}{2} < x < \dfrac{3(\sqrt{5} + \sqrt{3})}{2}.$

   Since

$$\frac{2}{3} < \sqrt{3} - 1 \Leftrightarrow \frac{4}{9} < 4 - 2\sqrt{3} \Longleftrightarrow 9\sqrt{3} < 16 \Leftrightarrow \sqrt{243} < \sqrt{256},$$

$$\sqrt{3} - 1 < 1 \Leftrightarrow \sqrt{3} < 2 = \sqrt{4},$$

$$\frac{10}{3} < \sqrt{5} + \sqrt{3} \Leftrightarrow \frac{100}{9} < 8 + 2\sqrt{15} \Leftrightarrow 14 < 9\sqrt{15} \Leftrightarrow \sqrt{196} < \sqrt{1215},$$

$$\sqrt{5} + \sqrt{3} < 4 \Leftrightarrow 8 + 2\sqrt{15} < 16 \Leftrightarrow \sqrt{15} < 4 \Leftrightarrow \sqrt{15} < \sqrt{16},$$

$$\therefore 1 < \frac{3(\sqrt{3} - 1)}{2} < \frac{3}{2}, \quad 5 < \frac{3(\sqrt{5} + \sqrt{3})}{2} < 6,$$

   $\therefore x$ may be 2, 3, 4, 5, the answer is (C).

9. Let $A = \dfrac{1}{1 - \sqrt[4]{5}} + \dfrac{1}{1 + \sqrt[4]{5}} + \dfrac{2}{1 + \sqrt{5}}$, then

$$A = \left(\frac{1}{1 - \sqrt[4]{5}} + \frac{1}{1 + \sqrt[4]{5}}\right) + \frac{2}{1 + \sqrt{5}} = \frac{2}{1 - \sqrt{5}} + \frac{2}{1 + \sqrt{5}} = \frac{2 \times 2}{1 - 5} = -1.$$

10. Since $U, V, W > 0$, It is sufficient to compare $U^2, V^2, W^2$. From

$$\begin{aligned} U^2 - V^2 &= (\sqrt{ab} + \sqrt{cd})^2 - (\sqrt{ac} + \sqrt{bd})^2 = ab + cd - ac - bd \\ &= (a - d)(b - c) > 0. \\ V^2 - W^2 &= (\sqrt{ac} + \sqrt{bd})^2 - (\sqrt{ad} + \sqrt{bc})^2 = ac + bd - ad - bc \\ &= (a - b)(c - d) > 0. \end{aligned}$$

   Therefore $U^2 > V^2 > W^2$, i.e. $W < V < U$.

## Testing Questions (16-B)

1. Considering negative number cannot be under square root sign, we find $|a| = 3$ i.e. $a = \pm 3$. Further, $3 - a$ appears in denominator implies $a \neq 3$, so $a = -3$. Thus

$$x = \left( \frac{(-2)(-3)}{4-3} \right)^{1993} = 6^{1993}.$$

Thus, the units digit of $x$ is 6 since any positive integer power of 6 always has units digit 6.

2. Let $\sqrt[3]{3} = x$, $\sqrt[3]{2} = y$, then the given expression becomes

$$x \left( \frac{y^2}{x^2} - \frac{y}{x^2} + \frac{1}{x^2} \right)^{-1} = x \left( \frac{y^2 - y + 1}{x^2} \right)^{-1} = x \cdot \frac{x^2}{y^2 - y + 1}$$

$$= \frac{x^3}{y^2 - y + 1} = \frac{x^3(y+1)}{y^3 + 1} = \frac{3(y+1)}{2+1} = y + 1 = \sqrt[3]{2} + 1.$$

3. More general, we calculate $A = \sqrt{\dfrac{n(n+1)(n+2)(n+3)+1}{4}}$.

$$A = \tfrac{1}{2}\sqrt{(n^2+3n)(n^2+3n+2)+1} = \tfrac{1}{2}\sqrt{(n^2+3n+1)^2 - 1 + 1}$$
$$= \frac{n^2 + 3n + 1}{2}.$$

Now $n = 1998$, so $A = \dfrac{3997999}{2} = 1998999.5$.

4. From $(\sqrt[3]{2} - 1)a = (\sqrt[3]{2})^3 - 1 = 1$, $a = \dfrac{1}{\sqrt[3]{2} - 1}$, so that

$$a^2 = \frac{1}{(\sqrt[3]{2} - 1)^2} = \frac{1}{\sqrt[3]{4} - 2\sqrt[3]{2} + 1}, a^3 = \frac{1}{2 - 3\sqrt[3]{4} + 3\sqrt[3]{2} - 1}.$$

Thus,

$$\frac{3}{a} + \frac{3}{a^2} + \frac{1}{a^3}$$
$$= 3(\sqrt[3]{2} - 1) + 3(\sqrt[3]{4} - 2\sqrt[3]{2} + 1) + 1 - 3\sqrt[3]{4} + 3\sqrt[3]{2}$$
$$= 1.$$

5. We find $[M]$ first. Let $A = \sqrt{13} + \sqrt{11}$, $B = \sqrt{13} - \sqrt{11}$. Then $M = A^6$ and $A + B = 2\sqrt{13}$, $AB = 2$, so

$$A^2 + B^2 = (A+B)^2 - 2AB = 52 - 4 = 48.$$

Since $M = A^6$, for finding $[M]$, we now consider $A^6 + B^6$:

$$A^6 + B^6 = (A^2)^3 + (B^2)^3 = (A^2 + B^2)(A^4 - A^2 B^2 + B^4)$$
$$= (A^2 + B^2)[(A^2 + B^2)^2 - 3(AB)^2]$$

which is an integer. $0 < B = \sqrt{13} - \sqrt{11} < 4 - 3 = 1$ yields $0 < B^6 < 1$, so $[M] = A^6 + B^6 - 1$ and

$$P = M - [M] = A^6 - (A^6 + B^6 - 1) = 1 - B^6, \text{ or } 1 - P = B^6.$$

Thus,
$$M(1 - P) = A^6 B^6 = (AB)^6 = 2^6 = 64.$$

# Solutions to Testing Questions    17

## Testing Questions    (17-A)

1.  $\sqrt{12 - 4\sqrt{5}} = \sqrt{(\sqrt{10} - \sqrt{2})^2} = \sqrt{10} - \sqrt{2}.$

2.  $\sqrt{2 + \sqrt{3}} + \sqrt{2 - \sqrt{3}} = \dfrac{1}{\sqrt{2}}[(\sqrt{3} + 1) + (\sqrt{3} - 1)] = \sqrt{6}.$

3.  $\sqrt{14 + 6\sqrt{5}} - \sqrt{14 - 6\sqrt{5}} = 3 + \sqrt{5} - (3 - \sqrt{5}) = 2\sqrt{5}.$

4.  Let $a = \sqrt{8 + \sqrt{63}}, b = \sqrt{8 - \sqrt{63}}$, then $a^2 + b^2 = 16, ab = 1$, so
$$(a - b)^2 = 16 - 2 = 14 \Longrightarrow a - b = \sqrt{14}.$$

5.  Let $a = \sqrt{4 + \sqrt{7}}, b = \sqrt{4 - \sqrt{7}}$, then $a^2 + b^2 = 8, ab = 3$, so
$$(a + b)^2 = 8 + 6 = 14 \Longrightarrow a + b = \sqrt{14}.$$

6.  $\sqrt{7 - \sqrt{15} - \sqrt{16 - 2\sqrt{15}}} = \sqrt{7 - \sqrt{15} - \sqrt{15} + 1} = \sqrt{5} - \sqrt{3}.$

7.  Since $xy = \dfrac{1}{4}[(x + y)^2 - (x - y)^2]$, so
$$xy = \dfrac{1}{4}[(3\sqrt{5} - \sqrt{2}) - (3\sqrt{2} - \sqrt{5})] = \sqrt{5} - \sqrt{2}.$$

8. Let $\sqrt{16 + 2(2 + \sqrt{5})(2 + \sqrt{7})} = a + b\sqrt{5} + c\sqrt{7}$, where $a, b, c > 0$. By taking squares to both sides, then

$$16 + 4\sqrt{5} + 4\sqrt{7} + 2\sqrt{35} = a^2 + 5b^2 + 7c^2 + 2ab\sqrt{5} + 2ac\sqrt{7} + 2bc\sqrt{35},$$

$$\therefore a^2 + 5b^2 + 7c^2 = 16, \quad ab = 2, \quad ac = 2, \quad bc = 1,$$
$$\therefore a^2 = 4, \quad \text{i.e. } a = 2, b = c = 1.$$

Thus, $\sqrt{8 + 2(2 + \sqrt{5})(2 + \sqrt{7})} = 2 + \sqrt{5} + \sqrt{7}$.

9. Let $A = \sqrt{a + 3 + 4\sqrt{a - 1}} + \sqrt{a + 3 - 4\sqrt{a - 1}} = \sqrt{(\sqrt{a - 1} + 2)^2}$ $+ \sqrt{(\sqrt{a - 1} - 2)^2}$, then $A = 2 + \sqrt{a - 1} + |\sqrt{a - 1} - 2|$.

It's clear that $a \geq 1$. When $\sqrt{a - 1} < 2$ i.e. $1 \leq a < 5$, then $A = 2 + \sqrt{a - 1} + 2 - \sqrt{a - 1} = 4$. When $5 \leq a$, then

$$A = \sqrt{a - 1} + 2 + \sqrt{a - 1} - 2 = 2\sqrt{a - 1}.$$

Thus,

$$\sqrt{a + 3 + 4\sqrt{a - 1}} + \sqrt{a + 3 - 4\sqrt{a - 1}} = \begin{cases} 4, & \text{if } 1 \leq a < 5, \\ 2\sqrt{a - 1}, & \text{if } a \geq 5. \end{cases}$$

10. It is clear that $x = \dfrac{\sqrt{6} - \sqrt{30}}{\sqrt{2} - \sqrt{10}} - 2 = \sqrt{3} - 2$. On the other hand,

$$A = \frac{(\sqrt{\sqrt{3} + 1} - \sqrt{\sqrt{3} - 1})^2}{(\sqrt{3} + 1) - (\sqrt{3} - 1)} = \frac{2\sqrt{3} - 4}{2} = \sqrt{3} - 2.$$

Thus, $A$ is the root of the given equation for $x$.

## Testing Questions   (17-B)

1. It is clear that $ab \geq 0$, i.e. $a$ and $b$ have same signs. Let $A = \sqrt{2\sqrt{ab} - a - b}$, then

$A = \sqrt{-b}$ if $a = 0$ and $b < 0$; or $A = \sqrt{-a}$ if $b = 0$ and $a < 0$.

$A$ is not defined if $a > 0, b > 0$ since $2\sqrt{ab} - a - b = -(\sqrt{a} - \sqrt{b})^2 < 0$.

$A = \sqrt{(\sqrt{-a} + \sqrt{-b})^2} = \sqrt{-a} + \sqrt{-b}$ if $a < 0, b < 0$.

Thus, $A = \sqrt{-a} + \sqrt{-b}$ if $A$ is defined.

2.  $\sqrt[3]{\dfrac{(\sqrt{a-1}-\sqrt{a})^5}{\sqrt{a-1}+\sqrt{a}}} + \sqrt[3]{\dfrac{(\sqrt{a-1}+\sqrt{a})^5}{\sqrt{a}-\sqrt{a-1}}}$

$$= -\sqrt[3]{(\sqrt{a-1}-\sqrt{a})^6} + \sqrt[3]{(\sqrt{a-1}+\sqrt{a})^6}$$

$$= -(\sqrt{a-1}-\sqrt{a})^2 + (\sqrt{a-1}+\sqrt{a})^2 = 4\sqrt{a(a-1)}.$$

3.  Since

$$1 + a^2 + \sqrt{1+a^2+a^4} = \frac{1}{2}[2 + 2a^2 + 2\sqrt{(a^2+1)^2 - a^2}]$$

$$= \frac{1}{2}[(a^2+a+1) + 2\sqrt{(a^2+a+1)(a^2-a+1)} + (a^2-a+1)]$$

$$= \frac{1}{2}(\sqrt{a^2+a+1} + \sqrt{a^2-a+1})^2,$$

therefore

$$\sqrt{1 + a^2 + \sqrt{1+a^2+a^4}} = \frac{\sqrt{a^2+a+1} + \sqrt{a^2-a+1}}{\sqrt{2}}$$

$$= \frac{\sqrt{2}(\sqrt{a^2+a+1} + \sqrt{a^2-a+1})}{2}.$$

4.  Let $\sqrt{2x-5} = y$, then $y \geq 0$ and $x = \dfrac{(y^2+5)}{2}$, so that

$$\sqrt{x+2+3\sqrt{2x-5}} - \sqrt{x-2+\sqrt{2x-5}}$$

$$= \sqrt{\frac{1}{2}(y^2+5) + 2 + 3y} - \sqrt{\frac{1}{2}(y^2+5) - 2 + y}$$

$$= \frac{1}{\sqrt{2}}\sqrt{y^2+6y+9} - \frac{1}{\sqrt{2}}\sqrt{y^2+2y+1}$$

$$= \frac{1}{\sqrt{2}}(y+3) - \frac{1}{\sqrt{2}}(y+1) = \frac{2}{\sqrt{2}} = \sqrt{2}.$$

5.  $\sqrt{x} = \sqrt{a} - \dfrac{1}{\sqrt{a}}$ yields $x = a + \dfrac{1}{a} - 2$, so $a + \dfrac{1}{a} = x + 2$. Since

$\dfrac{a-1}{\sqrt{a}} = \sqrt{x} \geq 0$, so $a \geq 1$, hence $a - \dfrac{1}{a} \geq 0$. Thus,

$$\sqrt{x^2+4x} = \sqrt{(x+2)^2 - 4} = \sqrt{(a+\frac{1}{a})^2 - 4} = \sqrt{(a-\frac{1}{a})^2} = a - \frac{1}{a},$$

it yields that

$$\frac{x+2+\sqrt{x^2+4x}}{x+2-\sqrt{x^2+4x}} = \frac{(a+\frac{1}{a}) + (a-\frac{1}{a})}{(a+\frac{1}{a}) - (a-\frac{1}{a})} = a^2.$$

6. $\dfrac{1}{\sqrt{17 - 12\sqrt{2}}} = \dfrac{1}{\sqrt{(3 - \sqrt{8})^2}} = \dfrac{1}{3 - \sqrt{8}} = 3 + \sqrt{8}$, so

$$3 + \sqrt{4} < 3 + \sqrt{8} < 3 + \sqrt{9} = 6, \text{ i.e. } 5 < 3 + \sqrt{8} < 6.$$

Since $5.5 = 3 + 2.5 = 3 + \sqrt{6.25} < 3 + \sqrt{8}$, the nearest integer is 6.

## Solutions to Testing Questions  18

### Testing Questions  (18-A)

1. The division with remainder $2007 = nq + 7$ implies $nq = 2000$, where $q$ is the quotient. Since $nq = 2000 = 2^4 \cdot 5^3$ has $(4 + 1) \cdot (3 + 1) = 20$ positive divisors and among them only $1, 2, 4, 5$ are less than 7, so there are $20 - 4 = 16$ divisors can be taken as $n$. Thus, the number of desired divisor $n$ is 16.

2. $123456789^4 \equiv 789^4 \equiv 5^4 \equiv 25^2 \equiv 1 \pmod{8}$, the remainder is 1.

3. Since $2222 \equiv 3 \pmod 7$ and $3^6 \equiv 1 \pmod 7$, it follows that
$$2222^{5555} \equiv 3^{6 \times 925} \cdot 3^5 \equiv 243 \equiv 5 \pmod 7.$$

Since $5555 \equiv 4 \pmod 7$ and $4^3 \equiv 1 \pmod 7$, it follows that
$$5555^{2222} \equiv 4^{2222} \equiv 4^{3 \times 740} \cdot 4^2 \equiv 16 \equiv 2 \pmod 7.$$

Thus, $2222^{5555} + 5555^{2222} \equiv 5 + 2 \equiv 0 \pmod 7$, i.e.
$$7 \mid (2222^{5555} + 5555^{2222}).$$

4. Since $47^{37^{27}} \equiv 3^{37^{27}} \pmod{11}$ and $3^5 = 243 \equiv 1 \pmod{11}$ and
$$37^{27} \equiv 2^{27} \equiv (2^4)^6 \cdot 2^3 \equiv 3 \pmod 5,$$

it follows that $37^{27} = 5k + 3$ for some positive integer $k$ and

$$47^{37^{27}} \equiv 3^{37^{27}} \equiv 3^{5k+3} \equiv (3^5)^k \cdot 27 \equiv 5 \pmod{11}.$$

Thus, the remainder is 5.

5. Since $9 \equiv -2 \pmod{11}$ and $2^{10} = 1024 = 1023 + 1 \equiv 1 \pmod{11}$,
$$9^{1990} \equiv (-2)^{1990} \equiv 2^{1990} \equiv (2^{10})^{199} \equiv 1 \pmod{11},$$

the remainder is 1.

6.  It is clear that $n$ is odd since it is the product of odd numbers. Let $x$ be the last three digits in that order, then $n \equiv x \pmod{1000}$. Since $15, 35, 55$ are three numbers in the product, so $n$ is divisible by $125$, hence $x$ is divisible by $125$. Thus, the possible values of $x$ are $125, 375, 625, 875$ only.

    On the other hand, $1000 \mid (n - x) \Leftrightarrow 8 \mid (n - x)$, so $n \equiv x \pmod 8$. For getting the remainder of $n$ modulo 8, we find that

    $$n = (3)(4 \cdot 1 + 3)(4 \cdot 2 + 3)(4 \cdot 3 + 3) \cdots (4 \cdot 499 + 3)(4 \cdot 500 + 3)$$
    $$\equiv \underbrace{(3 \cdot 7) \cdot (3 \cdot 7) \cdots (3 \cdot 7)}_{\text{250 pairs of brackets}} \cdot 3 \equiv \underbrace{5 \cdot 5 \cdots 5}_{\text{250 of 5}} \cdot 3 \quad \pmod 8$$
    $$\equiv \underbrace{1 \cdot 1 \cdots 1}_{125} \cdot 3 \equiv 3 \quad \pmod 8.$$

    Among $125, 375, 625, 875$, only $875$ has remainder 3 modulo 8, so $x = 875$, i.e. the last three digits of $n$ is $875$.

7.  Let $n$ be the solution. $n \equiv 1 \pmod 5$ implies $n = 5k + 1$ for some non-negative integer $k$;

    Then $n \equiv 2 \pmod 7$ implies $5k \equiv 1 \pmod 7$, so minimum $k$ is 3, i.e. $n = 16$ is the minimum $n$ satisfying the first two equations.

    Then $n = 16 + 35m$ is the general form of $n$. The requirement $n \equiv 3 \pmod 9$ yields $16 + 35m \equiv 3 \pmod 9$, and its minimum solution is $m = 4$, so $n = 156 + 315p$ for some $p$.

    Finally, from the fourth requirement $n \equiv 4 \pmod{11}$, we obtain $2 + 315p \equiv 4 \pmod{11}$, so $7p \equiv 2 \pmod{11}$, i.e. $p = 5$. Thus,

    $$n = 156 + 315 \times 5 = 1731.$$

8.  (a)  When $n = 3k$ where $k$ is a positive integer, then

    $$2^n - 1 \equiv (2^3)^k - 1 \equiv 1^k - 1 \equiv 0 \quad \pmod 7,$$

    so each multiple of 3 is a solution.

    When $n = 3k + r$, where $r = 1$ or 2, then

    $$2^n - 1 = (2^3)^k \cdot 2^r - 1 \equiv 2^r - 1 \equiv \begin{cases} 1 & \text{if } r = 1 \\ 3 & \text{if } r = 2. \end{cases}$$

    so $n = 3k$ for some positive integer $k$ is the necessary and sufficient condition for $7 \mid (2^n - 1)$.

    (b)  From (a) it is obtained that $2^n \equiv 0, 1$ or $3 \pmod 7$, so $2^n + 1 \not\equiv 0 \pmod 7$ for any positive integer $n$.

9. Since $3^4 \equiv 1 \pmod{10}$ and $7^4 = 2401 \equiv 1 \pmod{10}$,

$$3^{1999} = (3^4)^{499} \cdot 3^3 \equiv 7 \quad \pmod{10},$$
$$7^{2000} = (7^4)^{500} \equiv 1 \quad \pmod{10},$$
$$17^{2001} \equiv 7^{2001} \equiv 7^{2000} \cdot 7 \equiv 7 \quad \pmod{10},$$

so $3^{1999} \times 7^{2000} \times 17^{2001} \equiv 7 \cdot 1 \cdot 7 \equiv 9 \pmod{10}$, i.e. the solution is (E).

10. From $2^{10} = 1024 \equiv 24 \pmod{100}$ and $24^{2k} \equiv 576^k \equiv 76 \pmod{100}$,

$$2^{999} \equiv (2^{10})^{99} \cdot 2^9 \equiv (24)^{99} \cdot 2^9 \equiv 76 \cdot 24 \cdot 12 \quad \pmod{100}$$
$$\equiv 38 \cdot 24^2 \equiv 38 \cdot 76 \equiv 2888 \equiv 88 \quad \pmod{100}.$$

Therefore the last two digits of $2^{999}$ are 88.

### Testing Questions (18-B)

1. **Solution 1**  First of all we find the remainder of $14^{14^{14}}$ modulo 25.

$14^2 = 196 \equiv -4 \pmod{25} \Rightarrow (14)^5 \equiv (-4)^2 \cdot 14 \equiv 224 \equiv -1 \pmod{25}$,
so $(14)^{10} \equiv 1 \pmod{25}$. On the other hand,

$$14^2 = 196 \equiv 6 \pmod{10} \Rightarrow 14^{14} \equiv 6^7 \equiv 6 \pmod{10},$$

so $14^{14} = 10t + 6$ for some positive integer $t$, hence

$$14^{14^{14}} = 14^{10t+6} = (14^{10})^t \cdot 14^5 \cdot 14 \equiv (1)(-1)(14) \equiv 11 \pmod{25}.$$

Since $14^{14^{14}} = (2 \cdot 7)^{2(5t+3)} = 4^{5t+3} \cdot 7^{10t+6}$ which is divisible by 4, so $14^{14^{14}} \equiv 0 \pmod 4$. Write $14^{14^{14}} = 25K + 11$ where $K$ is a positive integer, then

$$25K + 11 \equiv 0 \quad \pmod 4,$$
$$K - 1 \equiv 0 \quad \pmod 4,$$
$$K \equiv 1 \quad \pmod 4, \quad \text{i.e. } K = 4l + 1 \text{ for some } l \in \mathbb{N},$$
$$\therefore 14^{14^{14}} = 25(4l + 1) + 11 = 100l + 36.$$

Thus the last two digits of $14^{14^{14}}$ are 36.

**Solution 2**  Notice that $14^{14} \equiv (4^2)^7 \equiv 6 \pmod{10}$ and

$$14^{10} \equiv (196)^5 \equiv (-4)^5 \equiv -1024 \equiv 76 \pmod{100},$$

so

$$14^{14^{14}} \equiv 14^{10t+6} \equiv 76 \cdot (-4)^3 \equiv 76 \cdot 36 \equiv 36 \pmod{100}.$$

2. Let $A = (257^{33} + 46)^{26}$. From

$$257^{33} \equiv 7^{33} \equiv (49)^{16} \cdot 7 \equiv 7 \quad (\text{mod } 50) \text{ and } 46 \equiv -4 \quad (\text{mod } 50),$$

it follows that

$$A = (257^{33} + 46)^{26} \equiv (7 - 4)^{26} \equiv 3^{26} \quad (\text{mod } 50).$$

$3^5 = 243 \equiv -7 \ (\text{mod } 50)$ yields $3^{10} \equiv (-7)^2 \equiv -1 \ (\text{mod } 50)$, so

$$A \equiv 3^{26} \equiv (3^{10})^2 \cdot 3^5 \cdot 3 \equiv (1)(-7)(3) \equiv -21 \equiv 29 \quad (\text{mod } 50),$$

i.e. the remainder of $(257^{33} + 46)^{26}$ modulo 50 is 29.

3. $3^n$ ends with 003 is equivalent to that $3^{n-1}$ ends with 001, i.e. the units digit of $3^{n-1}$ is 1, hence $n - 1 = 4k$ for some natural number $k$. Since $125 \mid 3^{4k} - 1$ and

$$3^{4k} - 1 = 81^k - 1 = 80(81^{k-1} + 81^{k-2} + \cdots + 81 + 1),$$

so $25 \mid (81^{k-1} + 81^{k-2} + \cdots + 81 + 1)$. It yields $5 \mid (81^{k-1} + 81^{k-2} + \cdots + 81 + 1)$. From

$$(81^{k-1} + 81^{k-2} + \cdots + 81 + 1) \equiv \underbrace{1 + 1 + \cdots + 1}_{k} \equiv k \quad (\text{mod } 5),$$

it follows that $k = 5m$ for some positive integer $m$. Below we find the minimum value of $m$ recursively. From

$$
\begin{aligned}
81 &\equiv 81 \quad (\text{mod } 1000), \\
81^2 &\equiv 6561 \equiv 561 \quad (\text{mod } 1000), \\
81^4 &\equiv 561^2 \equiv (500 + 61)^2 \equiv 61^2 \equiv 721 \quad (\text{mod } 1000), \\
81^5 &\equiv 721 \cdot 81 \equiv 58401 \equiv 401 \quad (\text{mod } 1000),
\end{aligned}
$$

therefore for $m = 2, 3, 4, 5$,

$$
\begin{aligned}
(81^5)^2 - 1 &\equiv (400 + 1)^2 - 1 \equiv 800 \quad (\text{mod } 1000), \\
(8a^5)^3 - 1 &\equiv 801 \cdot 401 - 1 \equiv 321201 - 1 \equiv 200 \quad (\text{mod } 1000), \\
(81^5)^4 - 1 &\equiv (800 + 1)^2 - 1 \equiv 600 \quad (\text{mod } 1000), \\
(81^5)^5 - 1 &\equiv 601 \cdot 401 - 1 \equiv 241001 - 1 \equiv 0 \quad (\text{mod } 1000).
\end{aligned}
$$

Thus, $m_{\min} = 5$, hence $k_{\min} = 25$ and $n_{\min} = 4k_{\min} + 1 = 101$ is the minimum required value.

4. Since $a + b + c = a + b + 2a + 5b = 3(a + 2b)$, so $3 \mid (a + b + c)$.
   Now let $a \equiv r_1 \ (\text{mod } 3)$ and $b \equiv r_2 \ (\text{mod } 3)$, then $0 < r_1, r_2 \leq 2$.

(i)   Suppose that $r_1 \neq r_2$. When $r_1 = 1, r_2 = 2$, then

$$c = 2a + 5b = 2(3q_1 + 1) + 5(3q_2 + 2) = 3(2q_1 + 5q_2 + 4) \Rightarrow c \text{ is not prime.}$$

When $r_1 = 2, r_2 = 1$, then $c = 2(3q_1 + 2) + 5(3q_2 + 1) = 3(2q_1 + 5q_2 + 3)$ i.e. $c$ is not prime also. Thus, $r_1 = r_2 = r$.

(ii)   When $r_1 = r_2 = r$ ($r$ may be 1 or 2), then

$$a + b + c = 3(a + 2b) = 3(3q_1 + r + 6q_2 + 2r) = 9(q_1 + 2q_2 + r)$$

which is divisible by 9, so $9 \mid (a + b + c)$.

Below by two examples we indicate that 9 is the maximum possible value of $n$. Let $a_1 = 11, b_1 = 5, c_1 = 47$, then $a_1 + b_1 + c_1 = 63$, and $n$ is a factor of 63.

Let $a_2 = 13, b_2 = 7, c = 61$, then $a_2 + b_2 + c_2 = 81$, $n$ is a factor of 81 also. From $(63, 81) = 9 \geq n$, the conclusion is proven.

5.   Note that $n \cdot 2^n + 1 \equiv 0 \equiv 3 \pmod 3 \Leftrightarrow n \cdot 2^n \equiv 2 \pmod 3$.

(i)   For $n = 6k + 1$, where $k$ is any non-negative integer,

$$n \cdot 2^n = (6k + 1) \cdot 2^{6k+1} \equiv 2 \cdot 4^{3k} \equiv 2 \pmod 3.$$

(ii)   For $n = 6k + 2$, where $k$ is any non-negative integer,

$$n \cdot 2^n = (6k + 2) \cdot 2^{6k+2} \equiv 8 \cdot 4^{3k} \equiv 2 \pmod 3.$$

(iii)   For $n = 6k + 3$, where $k$ is any non-negative integer,

$$n \cdot 2^n = (6k + 3) \cdot 2^{6k+1} \equiv 0 \pmod 3.$$

(iv)   For $n = 6k + 4$, where $k$ is any non-negative integer,

$$n \cdot 2^n = (6k + 4) \cdot 2^{6k+4} \equiv 4^{3k+2} \equiv 1 \pmod 3.$$

(v)   For $n = 6k + 5$, where $k$ is any non-negative integer,

$$n \cdot 2^n = (6k + 5) \cdot 2^{6k+5} \equiv 4 \cdot 4^{3k+2} \equiv 1 \pmod 3.$$

(vi)   For $n = 6k + 6$, where $k$ is any non-negative integer,

$$n \cdot 2^n = (6k + 6) \cdot 2^{6k+6} \equiv 0 \pmod 3.$$

Thus, the solution set is all $n$ with the forms $6k + 1$ or $6k + 2, k = 0, 1, 2, \ldots$.

# Solutions to Testing Questions    19

## Testing Questions    (19-A)

1.  From the assumption there exists a positive integer $k$ such that

$$100a + 10b + c = 37k,$$

therefore

$$
\begin{aligned}
\overline{bca} &= 100b + 10c + a = 1000a + 100b + 10c - 999a \\
&= 10(100a + 10b + c) - 37 \cdot 27a = 37(10k - 27a),
\end{aligned}
$$

which is divisible by 37.

2.  Let $n$ digit number $A$ be the number obtained by deleting the first digit 6, then

$$
\begin{aligned}
25A &= 6 \times 10^n + A, \\
4A &= 10^n, \\
\therefore A &= 25 \times 10^{n-2},
\end{aligned}
$$

i.e. the original number is $625 \times 10^{n-2}$ (where $n \geq 2$).

3.  Use $a, b, c$ to denote the hundreds, tens and units digits of $x$ respectively, then $a + b + c = 21$ and

$$100a + 10b + c + 495 = 100c + 10b + a,$$

therefore

$$99(c - a) = 495,$$

$$c - a = 5,$$

$$\therefore c = a + 5,$$

hence the possible values of $a$ are $1, 2, 3, 4$. Then the condition $a + b + c = 21$ implies $2a + b = 16$, so $b = 2(8 - a) \leq 9$, $a = 4$. Thus, $c = 9$, $b = 8$, i.e. $x = 489$.

4.  Let the given numbers be

$$A_1, \quad A_2, \quad A_3, \quad A_4, \cdots,$$

where $A_n = 7 \underbrace{11 \cdots 11}_{n-1} 2 \underbrace{88 \cdots 88}_{n-1} 9$ for $n = 1, 2, 3, \ldots$, then

$$A_n = 7 \cdot 10^{2n} + \underbrace{11 \cdots 11}_{n-1} \cdot 10^{n+1} + 2 \cdot 10^n + 8 \cdot \underbrace{11 \cdots 11}_{n-1} \cdot 10 + 9$$

$$= \frac{1}{9}[63 \cdot 10^{2n} + (10^{n-1} - 1) \cdot 10^{n+1} + 18 \cdot 10^n + 8 \cdot (10^{n-1} - 1) \cdot 10 + 81]$$

$$= \frac{1}{9}[64 \cdot 10^{2n} - 10^{n+1} + 18 \cdot 10^n + 8 \cdot 10^n + 1]$$

$$= \frac{1}{9}[(8 \cdot 10^n)^2 + 2 \cdot 8 \cdot 10^n + 1] = \frac{1}{9}(8 \cdot 10^n + 1)^2$$

$$= \left(\frac{8(10^n - 1)}{3} + 3\right)^2 = (2\underbrace{66 \cdots 66}_{n-1}7)^2.$$

Thus the conclusion is proven.

5.  Let $n = 10x + y$, where $y$ is the last digit, and $x$ is a $m$ digit number. Then

$$5(10x + y) = 10^m y + x,$$
$$49x = (10^m - 5)y.$$

Since $1 \le y \le 9 < 49$, so $7 \mid (10^m - 5)$. The minimum value of $m$ such that $7 \mid (10^m - 5)$ is 5, so

$$49x = 99995y \Rightarrow 7x = 14258y. \quad \because 7 \nmid 14258, \quad \therefore y = 7, x = 14258.$$

Thus, the number is 142587.

6.  Let $n = 1000a + 100b + 10c + d$, where $a, b, c, d$ be the digits of $n$. From the assumption,

$$1000a + 100b + 10c + d + a + b + c + d = 2001,$$
$$1001a + 101b + 11c + 2d = 2001,$$

so $a = 1$ and $101b + 11c + 2d = 1000$, which implies $b + c + 2d = 10(100 - 10b - c)$, therefore $10 \mid (b + c + 2d)$.

(i)  If $b + c + 2d = 10$, then $100 - 10b - c = 1$, i.e. $b = c = 9$ and $d = -4$, it is impossible.

(ii)  If $b + c + 2d = 20$, then $100 - 10b - c = 2$, i.e. $b = 9, c = 8$ and $2d = 3$, it is impossible.

(iii)  If $b + c + 2d = 30$, then $100 - 10b - c = 3$, i.e. $b = 9, c = 7$ and $d = 7$. By checking, it is certainly a solution.

Thus, $n = 1977$.

7.   Since $\underbrace{11\cdots11}_{1989 \text{ digits}} = \dfrac{1}{9}(10^{1989} - 1)$, so

$$N = \underbrace{11\cdots11}_{1989 \text{ digits}} \times \frac{1}{9}(10^{1989} - 1) = \frac{1}{9}[\underbrace{11\cdots11}_{1989 \text{ digits}} \times 10^{1989} - \underbrace{11\cdots11}_{1989 \text{ digits}}]$$

$$= \frac{1}{9} \times \underbrace{11\cdots11}_{1988 \text{ digits}} 0 \underbrace{88\cdots88}_{1988 \text{ digits}} 9 =$$

$$\underbrace{123456790\cdots123456790}_{220 \text{ blocks}} 12345678\,\underbrace{987654320\cdots987654320}_{220 \text{ blocks}}\,987654321$$

Thus, the sum $S$ of all digits of $N$ is given by

$$S = 221 \times 37 - 1 + 221 \times 44 + 1 = 221 \times 81 = 17901.$$

8.   Let the original four digit number be $n$. Then the new number $n'$ is $n + 5940$, so the units digit of $n$ is equal to the hundreds digit. Let $n = \overline{abcb} = 1000a + 100b + 10c + b$, then assumptions give

$$(1000c + 100b + 10a + b) - (1000a + 100b + 10c + b) = 5940,$$
$$990(c - a) = 5940,$$
$$\therefore c - a = 6.$$

When $n$ is minimum, then $a = 1, c = 7$, so $a + c = 8$. Since $a + b + c + b$ has a remainder 8 when it is divided by 9, so $2b$ is divisible by 9, i.e. $b = 9$ or 0. Since $n$ is odd, so $b = 9$. Thus, the four digit number is 1979.

9.   $y = 2(x + 1)$ implies that in the process of getting $y$ from $x$, the first digit and the last digit of $x$ have changed, so the first digit of $x$ is 2 and the last digit of $x$ is 5, but the first digit of $y$ is 5 and last digit of $y$ is 2. Write $x = \overline{2abc5}$, $y = \overline{5a'b'c'2}$, then

$$5(10^4) + 10^3a' + 10^2b' + 10c' + 2 = 2(2 \cdot 10^4 + 10^3a + 10^2b + 10c + 6),$$
$$9990 = 1000(2a - a') + 100(2b - b') + 10(2c - c').$$

If $a \neq a'$, then $2a > a'$ implies $a = 5, a' = 2$, so $2a - a' = 8$, then the right hand side must be less than the left hand side. Hence $a = a' = 9$. Thus, it follows that

$$990 = 100(2b - b') + 10(2c - c').$$

Similarly, $b = b' = 9$ and $90 = 10(2c - c')$, so $c = c' = 9$ also.

Thus, $x = 29995$.

10.  For the numbers $100, 200, \ldots, 900$, the ratios are all $100$. Below we show that $100$ is the maximum value.

For any three digit number $\overline{abc}$ which is different from $100, \ldots, 900$, at least one of $b, c$ is not $0$. So

$$a + b + c \geq a + 1.$$

Then $\overline{abc} = 100a + 10b + c < 100a + 100 = 100(a + 1)$ implies that

$$\frac{\overline{abc}}{a + b + c} < \frac{100(a + 1)}{a + 1} = 100.$$

The conclusion is proven.

## Testing Questions    (19-B)

1.  $4^3 < 100 \leq \overline{abc} \leq 999 < 10^3$ implies $5 \leq a + b + c \leq 9$.

When $a + b + c = 5$, $5^3 = 125 \neq (1 + 2 + 5)^3$, no solution.

When $a + b + c = 6$, $6^3 = 216 \neq (2 + 1 + 6)^3$, no solution.

When $a + b + c = 7$, $7^3 = 343 \neq (3 + 4 + 3)^3$, no solution.

When $a + b + c = 8$, $8^3 = 512 = n = (5 + 1 + 2)^3$, so $n = 512$ is a solution.

When $a + b + c = 9$, $9^3 = 729 \neq (7 + 2 + 9)^3$, no solution.

Thus, $n = 512$ is the unique solution.

2.  The relation $a^2 + b = c$ implies $(\underbrace{xx \cdots x}_{n})^2 + \underbrace{yy \cdots y}_{n} = \underbrace{zz \cdots z}_{2n}$, so

$$x^2(\underbrace{11 \cdots 1}_{n})^2 + y\underbrace{11 \cdots 1}_{n} = z\underbrace{11 \cdots 1}_{2n} = z(\underbrace{11 \cdots 1}_{n}) \cdot (1\underbrace{00 \cdots 0}_{n-1}1).$$

Hence

$$x^2(\underbrace{11 \cdots 1}_{n}) + y = z(1\underbrace{00 \cdots 0}_{n-1}1). \qquad (30.1)$$

Write $x^2 = 10u + v$, where $u, v$ are the tens digit and units digit of $x^2$ ($u$ may be zero), then (30.1) becomes

$$u \cdot \underbrace{11 \cdots 1}_{n}0 + v \cdot \underbrace{11 \cdots 1}_{n} + y = z \cdot 1\underbrace{00 \cdots 0}_{n-1}1. \qquad (30.2)$$

Now compare the tens digits on two sides of (30.2). On the right hand side, the tens digit is 0, but on the left hand side, it is $u + v + 1 - 10$ when $v + y \geq 10$, or $u + v - 10$ when $v + y < 10$. Thus, it follows that

$$u + v = 9 \ \text{ or } \ u + v = 10.$$

On the other hand, since $10u + v = x^2$ is a perfect square, so $x^2 = 09, 36, 81, 64$ i.e. $x = 3, 6, 9, 8$ only.

When $x = 3$, then $z = 1, y = 2$ $(n \geq 2)$.

When $x = 6$, then $z = 4, y = 8$ $(n \geq 2)$.

When $x = 8$, then $z = 7, y = 3$ $(n \geq 2)$.

When $x = 9$, then no solution.

3.   Let $n = 10a + b$, then $n' = 10b + a$. the given condition implies

$$10a + b = (10b + a)q + q, \ \ q < 10b + a,$$

i.e.

$$(10 - q)a - (10q - 1)b = q. \tag{30.3}$$

(i)   When $q \geq 5$, then (30.3) implies $45 \geq 5a \geq (10q-1)b+q \geq 49b+q$, so $b = 0$. By (30.3), $q = (10-q)a$, so $a = \dfrac{q}{10 - q}$, therefore $q = 5, a = 1$ or $q = 8, a = 4$, however, they do not satisfy the original equation. Thus, no solution for $q \geq 5$.

(ii)   When $q = 1$, (30.3) becomes $9(a - b) = 1$, no solution.

(iii)   When $q = 2$, (30.3) becomes $8a - 19b = 2$, i.e. $8(a-5) = 19(b-2)$, so $19 \mid a - 5$ implies $a = 5$, so $b = 2$.

(iv)   When $q = 3$, (30.3) becomes $7a - 29b = 3$, so $b \equiv 4 \pmod{7}$, i.e. $b = 4$. However, $7a = 119$ implies $a > 10$, a contradiction. So no solution if $q = 3$.

(v)   When $q = 4$, then (30.3) yields $6a - 39b = 4$, so $0 \equiv 1 \pmod 3$, a contradiction.

Thus, $a = 5, b = 2$ is the unique solution, $n = 52$.

4.   Let $x = 1000a + 100a + 10b + b = 1100a + 11b$ be the desired four digit number. Then $11 \mid x$. Since $x$ is a perfect square, so $11^2 \mid x$, i.e. $11 \mid (100a + b)$, which implies that $11 \mid (a + b)$. Since $1 \leq a + b \leq 18$, so $a + b = 11$. Thus,

$$x = 11(100a + b) = 11(99a + 11) = 11^2(9a + 1)$$

is a perfect square, hence $9a + 1$ so is also, which implies $a = 7$ and $x = 88^2 = 7744$.

5.  Let $x^2 > 0$ be the desired number, and $y$ its last two digits with $0 < y < 100$. The given condition implies that there is a positive integer $z$ such that $x^2 = 100z^2 + y$, so $(x - 10z)(x + 10z) = y$.

Let $x - 10z = u, x + 10z = v$. Then $x = \dfrac{u+v}{2}, z = \dfrac{v-u}{20}$. Since the value of $z$ increases 1 will let $x^2$ increase at least 200 without considering the change of $y$, it is needed to let $z$ be as large as possible for getting the maximal $x^2$. Therefore $v - u$ should be as large as possible.

$uv = y < 100$ implies that $1 \le u \le v < 100$, i.e.

$$z = \frac{v-u}{20} < \frac{100}{20} = 5,$$

so $z_{max} = 4$, i.e. $v - u = 80$ or $v = u + 80$. Since $u(u + 80) = y < 100$ implies $u < 2$, so $u = 1, v = 81$. Hence

$$x = \frac{81+1}{2} = 41, x^2 = 41^2 = 1681.$$

## Solutions to Testing Questions    20

### Testing Questions    (20-A)

1.  $(3k^2 + 3k - 4) + (7k^2 - 3k + 1) = 10k^2 - 3$ implies that its last digit is 7, so it must be not a perfect square for any natural number $k$, i.e. the required $k$ does not exist.

2.  Let $(x - 1)(x + 3)(x - 4)(x - 8) + m = n^2$. Then

$$\begin{aligned} n^2 &= [(x - 1)(x - 4)] \cdot [(x + 3)(x - 8)] + m \\ &= (x - 5x + 4)(x^2 - 5x - 24) + m \\ &= [(x - 5x - 10) + 14][(x - 5x - 10) - 14] + m \\ &= (x^2 - 5x - 10)^2 - 14^2 + m \\ &= (x^2 - 5x - 10)^2 - 196 + m. \end{aligned}$$

Thus, $m = 196$, the answer is (D).

3.  Let $n + 20 = a^2, n - 21 = b^2$, where $a, b$ are two natural numbers. Then

$$41 = a^2 - b^2 = (a - b)(a + b)$$

which implies $a - b = 1, a + b = 41$, so $a = 21, b = 20$ and $n = 21^2 - 20 = 421$.

4.  Let the 2009 consecutive positive integers be

$$x - 1004, x - 1003, \cdots x - 1, x, x + 1, \cdots, x + 1004.$$

Then their sum is $2009x$. Since it is a perfect square and $2009 = 41 \cdot 49 = 41 \cdot 7^2$, the minimum value of $x$ is 41, i.e. the minimum value of $x + 1004$ is 1045.

5.  Since $4^{27} + 4^{1000} + 4^x = 2^{54} + 2^{2000} + 2^{2x} = 2^{54}(1 + 2 \cdot 2^{1945} + 2^{2x-54})$, it is obvious that the right hand side is a perfect square if $2^{2x-54} = (2^{1945})^2$, i.e., $x - 27 = 1945, x = 1972$.

When $x > 1972$, then

$$(2^{x-27})^2 = 2^{2x-54} < 1 + 2 \cdot 2^{1945} + 2^{2x-54} < (2^{x-27} + 1)^2,$$

so $1 + 2 \cdot 2^{1945} + 2^{2x-54}$ is not a perfect square. Thus, the maximal required value of $x$ is 1972.

6.  Note that the following inequality holds:

$$(n^2 + n)^2 < n^4 + 2n^3 + 2n^2 + 2n + 1 < n^4 + 2n^3 + 3n^2 + 2n + 1 = (n^2 + n + 1)^2,$$

the conclusion is proven at once.

7.  We prove by contradiction. Suppose that the $D = n^2$ satisfies all the requirements. The units digit of $D$ is 5 implies that $n$ is too. Assume $n = 10a + 5$, then $D = (10a + 5)^2 = 100a(a + 1) + 25$, so the last two digits of $D$ are 25.

Since the last digit of $a(a + 1)$ is 0, 2 or 6, and 0, 2 are impossible, so the third digit of $D$ is 6, i.e. $D = 1000b + 625$ for some digit $b$. Thus, $5^3 \mid D$, hence $5^4 \mid D$ since its a perfect square. However, it implies that $5^4 \mid 1000b$, so $5 \mid b$ i.e. $b = 0$ or 5, a contradiction.

8.  Let $n = \overline{abc} + \overline{bca} + \overline{cab}$, then

$$\begin{aligned} n &= (100a + 10b + c) + (100b + 10c + a) + (100c + 10a + b) \\ &= 111(a + b + c) = 3 \cdot 37(a + b + c). \end{aligned}$$

If $n = m^2$, then $37 \mid m^2$ implies $37^2 \mid 3 \cdot 37(a + b + c)$, so

$$37 \mid (a + b + c).$$

However, $0 < a + b + c \le 9 + 9 + 9 = 27$, a contradiction.

9.  Since each perfect square has remainder 0, 1 or 4 modulo 8, by taking modulo 8 to both sides of the equation, the left hand side is 0, 1, 2, 4 or 5, whereas the right hand side is 6, so the equality is impossible.

10. From assumption, there exists integer $k$ such that $x^2 + y^2 - x = 2kxy$. Consider the quadratic equation in $y$

$$y^2 - 2kxy + (x^2 - x) = 0.$$

since it has an integer solution, so, by Viete Theorem, the other root is also an integer, and the discriminant $\Delta$ of the equation is a perfect square. Hence

$$\Delta = 4[k^2x^2 - (x^2 - x)] = 4x[(k^2 - 1)x + 1]$$

is a perfect square. Since $x$ and $(k^2 - 1)x + 1$ are relatively prime, $x$ and $(k^2 - 1)x + 1$ both are perfect square numbers, so $x$ is a perfect square.

## Testing Questions    (20-B)

1. Write $A = (m + n)^2 + 3m + n$. Then $(m + n)^2 < A < (m + n)^2 + 4(m + n) + 4 = (m + n + 2)^2$. Therefore

$$(m + n)^2 + 3m + n = (m + n + 1)^2 = m^2 + n^2 + 1 + 2mn + 2m + 2n,$$
$$\therefore m = n + 1,$$

Thus, the solutions for $(m, n)$ are $(2, 1), (3, 2), \cdots, (99, 98)$, i.e. there are 98 required pairs for $(m, n)$.

2. The positive divisors of $p^4$ are $1, p, p^2, p^3, p^4$, therefore $1 + p + p^2 + p^3 + p^4 = n^2$ for some positive integer $n$, so

$$(p^2 + p)(p^2 + 1) = n^2 - 1 = (n - 1)(n + 1).$$

For $p = 2$, we have $1 + p + p^2 + p^3 + p^4 = 31$ which is not a perfect square, so $p \geq 3$.

Suppose that $n - 1 < p^2 + 1$, then $n + 1 < p^2 + 3 \leq p^2 + p$, so that $n^2 - 1 < (p^2 + 1)(p^2 + p)$. Thus, $n - 1 \geq p^2 + 1$, i.e. $n \geq p^2 + 2$. Let $n = p^2 + d$, then $d \geq 2$ and $n + 1 = p^2 + d + 1$, hence

$$p + p^2 + p^3 + p^4 = (p^2 + d)^2 - 1 = p^4 + 2d \cdot p^2 + d^2 - 1,$$
$$p^3 - (2d - 1)p^2 + p = (d - 1)(d + 1),$$
$$p[p^2 - (2d - 1)p + 1] = (d - 1)(d + 1).$$

If $p \mid (d - 1)$, i.e. $d = 1 + kp$ for some positive integer $k$, then

$$\dot{0} < p^2 - (1 + 2kp)p + 1 = (1 - 2k)p^2 - (p - 1) < 0,$$

a contradiction. Therefore $p \mid d + 1$, i.e. $d = kp - 1$ for some positive integer $k$, and,

$$1 \le p^2 - (2kp - 3)p + 1 \Longrightarrow p - (2kp - 3) \ge 0 \Longrightarrow (2k - 1)p \le 3$$
$$\Longrightarrow k = 1, p = 3.$$

Thus, $p = 3$ is the unique solution.

3.  (i)  When $x \ge y$, then

$$x^2 < x^2 + y + 1 \le x^2 + x + 1 < (x + 1)^2,$$

so $x^2 + y + 1$ is not a perfect square.

(ii)  When $x < y$, then $y^2 < y^2 + 4x + 3 < y^2 + 4y + 4 = (y + 2)^2$, so $y^2 + 4x + 3$ is not a perfect square if $y^2 + 4x + 3 \ne (y + 1)^2$. When

$$y^2 + 4x + 3 = (y + 1)^2 = y^2 + 2y + 1,$$

then $y = 2x + 1$, so that $x^2 + y + 1 = x^2 + 2x + 2$. However,

$$(x + 1)^2 = x^2 + 2x + 1 < x^2 + 2x + 2 < x^2 + 4x + 4 = (x + 2)^2$$

indicates that $x^2 + y + 1 = x^2 + 2x + 2$ is not a perfect square. Thus, the conclusion is proven.

4.  Since $2 \cdot 5 - 1, 2 \cdot 13 - 1$ and $5 \cdot 13 - 1$ are all perfect squares, it is necessary to show that at least one of $2d - 1, 5d - 1$ and $13d - 1$ is not a perfect square.

(i)  When $d$ is even i.e. $d = 2m$ for some positive integer $m$, then $2d - 1 = 4m - 1$ is not a perfect square (since its remainder modulo 4 is 3).

(ii)  When $d = 4m + 3$ for some non-negative integer $m$, then

$$5d - 1 = 20m + 14 = 4(5m + 3) + 2 \equiv 2 \pmod{4},$$

so $5d - 1$ is not a perfect square.

(iii)  When $d = 4m + 1$ for some non-negative integer $m$, then

$$5d - 1 = 20m + 4 = 4(5m + 1), \quad 13d - 1 = 52m + 12 = 4(13m + 3).$$

In case that $m \equiv 1$ or $2 \pmod{4}$, then $5m + 1 \equiv 2$ or $3 \pmod{4}$ implies that $5m + 1$ is not a perfect square, so $5d - 1$ is not a perfect square.

In case that $m \equiv 0$ or $3 \pmod{4}$, then $13m + 3 \equiv 3$ or $2 \pmod{4}$ implies that $13m + 3$ is not a perfect square, so $13d - 1$ is not a perfect square.

Thus, in any case at least one of three numbers $2d - 1, 5d - 1, 13d - 1$ is not a perfect square.

5. It is needed to consider several possible cases as follows.

Let $A = 3^n + 2 \times 17^n$.

(i) When $n = 4m$ where $m$ is a non-negative integer, then

$$3^n + 2 \cdot 17^n = (3^4)^m + 2(17^4)^m = 81^m + 2 \cdot 83521^m \equiv 3 \pmod{10},$$

so $A$ is not a perfect square.

(ii) When $n = 4m + 1$ where $m$ is a non-negative integer, then

$$3^n + 2 \cdot 17^n = 3 \cdot 81^m + 34 \cdot 83521^m \equiv 7 \pmod{10},$$

so $A$ is not a perfect square.

(iii) When $n = 4m + 2$ where $m$ is a non-negative integer, then

$$3^n + 2 \cdot 17^n = 9 \cdot 81^m + 578 \cdot 83521^m \equiv 7 \pmod{10},$$

so $A$ is not a perfect square.

(iv) When $n = 4m + 3$ where $m$ is a non-negative integer, then

$$3^n + 2 \cdot 17^n = 27 \cdot 81^m + 9826 \cdot 83521^m \equiv 3 \pmod{10},$$

so $A$ is not a perfect square.

Thus, $A$ is never a perfect square.

## Solutions to Testing Questions 21

### Testing Questions (21-A)

1. When we partition the $n + 1$ integers according to the remainders modulo $n$, there are at most $n$ classes. By the pigeonhole principle, there must be one class containing at least 2 numbers. Then any two of them have a difference which is divisible by $n$.

2. Write each of the $n + 1$ given numbers in the form $2^m \cdot q$, where $q$ is an odd number and $m$ is a non-negative integer. Then $1 \leq q \leq 2n - 1$, i.e. there are at most $n$ different values for $q$.

   By partitioning the given $n + 1$ numbers into at most $n$ classes according to the value of $q$, then, by the pigeonhole principle, there must be one class containing at least 2 numbers. The larger one of them must be divisible by the smaller one since they have the same $q$.

3. Suppose that each of the given numbers is not divisible by 2009. Then 0 is not a value of the remainders of the given numbers when divided by 2009, i.e. the 2009 remainders can take at most 2008 distinct values. Therefore, by the pigeonhole principle, there must be at least two of them, say $\underbrace{111\cdots111}_{p \text{ digits}}$

   and $\underbrace{111\cdots111}_{q \text{ digits}}$, where $1 \le p < q \le 2009$, such that they have a same remainder, and hence their difference

   $$\underbrace{111\cdots111}_{q-p \text{ digits}} \times 1\underbrace{000\cdots000}_{p \text{ digits}}$$

   is divisible by 2009. Then $(10^p, 2009) = 1$ implies $2009 \mid \underbrace{111\cdots111}_{q-p \text{ digits}}$, a contradiction. Thus, the conclusion is proven.

4. Partition each side of the equilateral triangle into three equal parts, and by passing through these partition points introduce lines parallel to the sides of the triangle, then the triangle will be partitioned into 9 equal small equilateral triangles of area $1/9$ cm$^2$.

   By the pigeonhole principle, there must be three points inside or on the boundary of a small triangle, then its area is not greater than $\dfrac{1}{9}$ m$^2$.

5. First of all, use the remainder modulo 10, or the units digit to replace the original one. Next, we consider the following six "pigeonholes":

   $$\{0\}, \ \{5\}, \ \{1,9\}, \ \{2,8\}, \ \{3,7\}, \ \{4,6\}.$$

   (i) if two remainders are both 0 or 5, their difference of the two numbers is divisible by 10.

   (ii) if two remainders are one of $(1,9), (2,8), (3,7), (4,6)$, then the sum of the two numbers is divisible by 10.

   Since seven remainders put in above six pairs, there must be two to be in the same pair, so their sum or difference is divisible by 10.

6. The 28 cells in the first row are colored by three colors, , by the pigeonhole principle, there must be at least $\lfloor 27/3 \rfloor + 1 = 10$ of them being of the same color, say it is red. By exchanging the columns if necessary, we can assume that the 10 columns with red cells are the first 10 columns.

   If there are two red cells appeared in the first 10 columns of another row, then the conclusion is proven.

Otherwise, there is at most one red cell appeared in the first 10 columns of each of the second, third or fourth row, by exchanging the first 10 columns if necessary, we may assume that the red cells are all in the first three columns of the 2nd, 3rd and 4th row, so in the rectangular region $A$ of dimension $3 \times 7$, which are formed by the last three rows and the 4th column to the 10th column, each cell is colored by blue or yellow color only. Then in the first row of $A$, by the pigeonhole principle, there must be four cells with the same color, say it is blue, and for convenience, we can assume that they are in the first four columns of $A$.

If in another row of $A$ there are two blue cells appeared in the first four columns of $A$, the conclusion is proven also, otherwise, in each other row of $A$ there is at most one blue cell in the first four columns, then in each other row of $A$ there are at least 3 yellow cells in the first four columns.

Then, assuming that in the second row of $A$ the three yellow cells appeared in the first 3 columns, since in the first three cells of third row of $A$ at least two are yellow, a required rectangular region with four yellow corners is obtained.

Thus, in any possible case the conclusion is proven.

7. As shown in the right digram, let $PQ$ be such a stright line and $MN$ be the common midline of the two trapizia. Since the two trapizia have equal height, the ratio of their areas is equal to that of their midlines, that is,

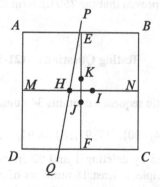

$MH/NH = 2/3$. Thus, $MH : MN = 2/5$. By symmetry, there are another three points $K, I, J$ on the two midlines of $ABCD$.

Each of the nine lines must pass through one of the four points, so, by the pigeonhole principle, there must be at least three of them passing through the same point.

8.  Partition the segment $OA$ into $n$ short segments of equal lengths $1/n$. By the pigeonhole principle, there must be two of the $n + 1$ points which are contained by a same short segment, so the distance between the two points is not greater than $1/n$.

9.  From the given condition, if the desired term exists, it must be at least a five digit number. Let $x_n$ be the last four digits of the $n$th term (if is is really less than 1000, then make it be a four digit number by adding pre-zeros, for example $204 = 0204$). Consider the following $100000001 = 10^8 + 1$ pairs

$$(x_1, x_2), \quad (x_2, x_3), \quad \ldots, (x_{100000001}, x_{100000002}).$$

Since the number of possible different two dimensional values of the pairs at most is $10^4 \times 10^4 = 10^8 < 10^8 + 1$, by the pigeonhole principle, in above $10^8 + 1$ pairs there must be two with same components, i.e. there exist $i, j$ with $1 \le i < j \le 10^8 + 1$ such that

$$x_i = x_j, \quad x_{i+1} = x_{j+1}.$$

Then $x_{i-1} = x_{i+1} - x_i = x_{j+1} - x_j = x_{j-1}$, so $(x_{i-1}, x_i) = (x_{j-1}, x_j)$. Continuing the process, it follows that

$$(x_1, x_2) = (x_{j-i+1}, x_{j-i}).$$

Then $x_1 = x_{j-i+1}$ implies that the $x_{j-i+1} = 0000$, i.e. the $j - i + 1$th term ends with at least four zeros.

In fact, it can be proven that the 7501th term has the desired property.

## Testing Questions   (21-B)

1.  The given arithmetic sequence contains 34 numbers in total. We consider the 16 pairs
$$\{4, 100\}, \quad \{7, 97\}, \quad \{10, 94\}, \quad \ldots, \quad \{49, 55\}$$

which are formed by deleting 1 and 52 from the sequence. Then by the pigeonhole principle, at least 18 numbers of $A$ is inside the 16 pairs, so at least there must be two pairs such that their components are all in $A$, i.e., there must be four numbers $a, b, c, d \in A$ such that

$$a + b = c + d = 104.$$

**Note:**   A stronger conclusion is proven actually here.

2. According to the units digit partition the 17 numbers into 5 classes: $C_0, C_1, C_2, C_3, C_4$.

   If each class contains at least one number, then the five numbers obtained by taking anyone from each class satisfy the requirement.

   If at least one class is empty set, then, by the pigeonhole principle, there must be one class containing at least $\lfloor 16/4 \rfloor + 1 = 5$ numbers. Any five numbers taking from this class have the same units digit, so their sum must be divisible by 5. The conclusion is proven also.

3. The set of 10 distinct two digit numbers has $2^{10} = 1024$ subsets totally (including the empty set and the set its self. For a proper subset, if use $S$ to denote the sum of all its elements, then $10 \le S \le 91 + 92 + \cdots + 99 = 855$. So the value of $S$ has $855 - 9 = 846$ choices.

   Since the 1022 proper subsets can take 846 possible different value of $S$, by the pigeonhole principle, there must be two subsets with equal value of $S$.

   Let $A$ and $B$ be two subsets with equal sum of elements, i.e. $S_A = S_B$. Then $A - \{A \cap B\}$ and $B - \{A \cap B\}$ also have equal sum of elements. Since $A \ne B$, so the two difference sets both are not empty.

4. Use 17 points $A_1, A_2, \ldots, A_{17}$ in the space to denote the 17 people, such that any three points are not collinear. Every two points are connected by a segment, and a segments are colored red, blue or white according to the two people discussed the first, second or third topic respectively. The question becomes to show that there must be a triangle with three sides of the same color.

   Among the 16 segments emitted from $A_1$, by the pigeonhole principle, there must be at least 6 with the same color. without loss of generality, one may assume that they are red, and their another end points are $A_2, A_3, A_4, A_5, A_6$ and $A_7$ respectively. If some of the segments joining the six points is red, then we have got a triangle with three red sides, the conclusion is proven.

   If all the segments joining these six points are of blue color or white color, then, from the result of Example 9, there must be a homochromatic triangle formed by these segments, the conclusion is also proven.

5. We prove the conclusion by contradiction. Suppose that the conclusion is not true, then the following property $P$ holds: the difference of any two members from a same country is not the number of a member from the same country.

   By the pigeonhole principle, there must be a country $A$ such that the number of members coming from this country $A$ is not less than $[1977/6] + 1 = 330$.

Let $m_1$ be the maximum number of the members from the country $A$. By using $m_1$ minus each of the numbers of rest members from the country $A$, not less than 329 differences can be obtain, and they must be the numbers of members not from the country $A$.

Since $[328/5] + 1 = 66$, by the pigeonhole principle again, there must be a country $B$ such that the number of members coming from this country $B$ is not less than 66. If $m_2$ is the maximum number of the 66 members from the country $B$, similarly at least 65 differences are obtained, and they must be neither the numbers of the members from the country $B$ nor that from the country $A$ (since $(m_1 - a_1) - (m_1 - a_2) = a_2 - a_1$ for any numbers $a_1, a_2$ from the country $A$). So these 65 differences must be numbers of the members from the rest four countries.

From $[64/4] + 1 = 17$, there must be a country $C$ such that the number of members coming from this country $C$ is not less than 17. So there must be at least 16 numbers of the members from the rest three countries. By using the pigeonhole principle once again, there must be a country $D$ with not less than 6 members. They yield at least 5 numbers of the members from the rest two countries. So there must be a country $E$ with not less than 3 members. Their three numbers yield two numbers of the members from the sixth country $F$. However, the difference of the two numbers yields a number which is not a number of all the six countries, a contradiction! The conclusion thus is proven.

# Solutions to Testing Questions    22

## Testing Questions    (22-A)

1.  Since

$$(\sqrt{n} + \sqrt{n + 1})^2 = 2n + 1 + 2\sqrt{n(n + 1)} > 2n + 1 + 2\sqrt{n^2} = 4n + 1$$

and

$$(\sqrt{n} + \sqrt{n + 1})^2 = 2n + 1 + 2\sqrt{n(n + 1)} < 2n + 1 + 2\sqrt{(n + 1)^2} = 4n + 3,$$

letting $k = [\sqrt{4n + 2}]$, then $k \le \sqrt{4n + 2} < k + 1$, i.e. $k^2 \le 4n + 2 < (k + 1)^2$. Since $4n + 2$ and $4n + 3$ both are not perfect square numbers (a perfect square number cannot have a remainder 2 or 3 when divided by 4), therefore
$$k^2 < 4n + 2 < 4n + 3 < (k + 1)^2,$$

then $k^2 \leq 4n + 1 < (\sqrt{n} + \sqrt{n+1})^2 < 4n + 3 \leq (k+1)^2$, i.e. $k < \sqrt{n} + \sqrt{n+1} < (k+1)$. Thus, $[\sqrt{n} + \sqrt{n+1}] = k = [\sqrt{4n+2}]$.

2.  Since the left hand side of the equation is an integer, therefore $\{x\} - 1$ is an integer. Then $0 \leq \{x\} < 1$ implies $\{x\} = 0$ i.e. $x = [x]$. Thus,

$$x^3 + x^2 + x = -1,$$
$$(x^3 + x^2) + (x + 1) = 0,$$
$$(x + 1)(x^2 + 1) = 0,$$

therefore $x = -1$.

3.  (CMO/1975) Solve equation $[x]^2 = \{x\} \cdot x$.

Let $[x] = n$, $t = \{x\}$, then $n^2 = t(n + t) \geq 0$. $0 \leq t < 1$ implies $n \geq 0$. Since $n^2 = t(n + t) < n + 1$, so $n = 0$ or 1. If $n = 0$ then $t = 0$, i.e. $x = 0$. If $n = 1$, then

$$1 = t(1 + t),$$
$$t^2 + t - 1 = 0,$$

$$t = \frac{-1 + \sqrt{5}}{2}, \quad \therefore \ x = n + t = \frac{1 + \sqrt{5}}{2}.$$

Thus, $x = 0$ or $x = \dfrac{1 + \sqrt{5}}{2}$.

4.  Let $x$ be a solution with $\lfloor x \rfloor = n$. Then the given equation becomes

$$x^2 + 7 = 8n.$$

So $n > 0$. The relation $n \leq x < n + 1$ yields

$$n^2 + 7 \leq x^2 + 7 = 8n < (n + 1)^2 + 7 = n^2 + 2n + 8,$$
$$\therefore n^2 - 8n + 7 \leq 0 \quad \text{and} \quad n^2 - 6n + 8 > 0.$$

the solutions for $n$ are $1 \leq n < 2$ or $4 \leq n < 7$, i.e. $n = 1, 5, 6, 7$.

Correspondingly, $x^2 + 7 = 8, 40, 48, 56$, so $x = 1, \sqrt{33}, \sqrt{41}, 7$ since $x > 0$.

5.  It is obvious that $x = n$ is a required root. Let $1 \leq x < n$ be another root of the given equation. Write $m = [x]$, $t = \{x\}$, then $x = m + t$ and the equation becomes

$$(m + t)^2 - [(m + t)^2] = t^2,$$
$$2mt - [2mt + t^2] = 0,$$

therefore $2mt$ is an non-negative integer, i.e. $t = 0, \frac{1}{2m}, \frac{2}{2m}, \cdots, \frac{2m-1}{2m}$ for $m = 1, 2, \cdots, n-1$. Thus, the number of required roots is

$$2[1 + 2 + 3 + \cdots + (n-1)] + 1 = (n-1)n + 1 = n^2 - n + 1.$$

6.   Let $n = 2x - \dfrac{1}{2}$, then $n \in \mathbb{Z}$ and $x = \frac{1}{2}n + \frac{1}{4}$, therefore

$$\left\lfloor \frac{3}{2}n + \frac{7}{4} \right\rfloor = n \Rightarrow n \le \frac{3}{2}n + \frac{7}{4} < n + 1 \Rightarrow -\frac{7}{2} \le n < -\frac{3}{2},$$

therefore $n = -3$ or $n = -2$. When $n = -3$, then $x = -\dfrac{3}{2} + \dfrac{1}{4} = -\dfrac{5}{4}$; when $n = -2$, then $x = -1 + \dfrac{1}{4} = -\dfrac{3}{4}$. Thus, the sum of roots is $-2$.

7.   $\dfrac{10^n}{x} - 1 < \left\lfloor \dfrac{10^n}{x} \right\rfloor \le \dfrac{10^n}{x}$ implies $\dfrac{10^n}{x} - 1 < 1989 \le \dfrac{10^n}{x}$, i.e. $\dfrac{10^n}{1990} < x \le \dfrac{10^n}{1989}$, so

$$10^n \cdot 0.00050251256\ldots < x \le 10^n \cdot 0.00050276520\ldots,$$

So only for $n \ge 7$ the difference of the two decimals is greater than 1, and it follows when $n = 7$ that

$$5025 < x \le 5027,$$

and $x = 5026$ or $5027$.

8.   Let the sum be $N$ and $x = n + t$ where $n \in \mathbb{Z}_0^+$ and $0 \le t < 1$. Then

$$N = 20n + \lfloor 2t \rfloor + \lfloor 4t \rfloor + \lfloor 6t \rfloor + \lfloor 8t \rfloor \le 1000,$$
$$\therefore 0 \le n \le 50.$$

For any fixed $0 \le n \le 49$, the the left hand side has different values at $t = \frac{r}{8}$ where $r = 0, 1, 2, 3, \cdots, 7$ and $t = \frac{s}{6}$ where $s = 1, 2, 4, 5$. Therefore there are totally 12 different values of $N$ for each $n = 1, \cdots, 49$, but there are a total of 11 different values of $N$ when $n = 0$ (since $t$ cannot take value 0), and $n = 50$, $t = 0$ means $N = 1000$ is also a required value, therefore $N$ takes a total of

$$50 \times 12 = 600 \text{ different positive integral values not greater than 1000.}$$

9. $(k+1)^2 - k^2 = 2k+1$ indicates that $\left[\dfrac{(k+1)^2}{1980}\right] > \left[\dfrac{k^2}{1980}\right]$ when $2k+1 >$ 1980 i.e. $k \geq 990$. Therefore $\left[\dfrac{k^2}{1980}\right]$, $k = 990, 991, \cdots, 1980$ are 991 different values.

For the sequence

$$\left[\frac{1^2}{1980}\right], \left[\frac{2^2}{1980}\right], \left[\frac{3^2}{1980}\right], \cdots, \left[\frac{990^2}{1980}\right],$$

since any two consecutive terms have difference 0 or 1 only, and $\left[\dfrac{990^2}{1980}\right] =$ 495, the sequence must take $495 + 1 = 496$ different values. Thus, altogether the whole sequence takes $496 + (991 - 1) = 1486$ different values.

10 Let $n$ be a required positive integer satisfying all requirements in question. Write $k = [\sqrt{n}]$, then $1 \leq k \leq 10^2$, and

$$k^2 \leq n < (k+1)^2, \qquad \text{or equivalently,} \qquad k^2 \leq n \leq k^2 + 2k,$$

i.e. $n = k^2 + r$ where $r$ is a non-negative integer with $0 \leq r \leq 2k$. The requirement $k \mid k^2 + r$ implies $r$ can take $0, k$ and $2k$ for $1 \leq k \leq 99$ and $r = 0$ for $k = 100$ only. Thus, totally there are

$$3 \times 99 + 1 = 298$$

desired $n$.

## Testing Questions (22-B)

1. By definition of integer part, it must be true that

$$\left\lfloor \sqrt{\sqrt{x}} \right\rfloor \leq \sqrt{\sqrt{x}} < \left\lfloor \sqrt{\sqrt{x}} \right\rfloor + 1.$$

Let $\left\lfloor \sqrt{\sqrt{x}} \right\rfloor = n$, then $n^4 \leq x < (n+1)^4$, so $n^2 \leq \sqrt{x} < (n+1)^2$ which implies that

$$n^2 \leq \lfloor \sqrt{x} \rfloor < (n+1)^2,$$

so $n \leq \sqrt{\lfloor \sqrt{x} \rfloor} < n+1$, and it implies that

$$\left\lfloor \sqrt{\lfloor \sqrt{x} \rfloor} \right\rfloor = n = \left\lfloor \sqrt{\sqrt{x}} \right\rfloor,$$

i.e., the given equality must be true.

2.  In the Q1 of (22-A), we have proven that $\sqrt{n} + \sqrt{n+1} > \sqrt{4n+1}$ and that

$$\lfloor \sqrt{n} + \sqrt{n+1} \rfloor = \lfloor \sqrt{4n+2} \rfloor = \lfloor \sqrt{4n+3} \rfloor.$$

It suffices to show that $\lfloor \sqrt{4n+1} \rfloor = \lfloor \sqrt{4n+3} \rfloor$. Let $\lfloor \sqrt{4n+1} \rfloor = k$, then $k \leq \sqrt{4n+1} < k+1$, so $k^2 \leq 4n+1 < (k+1)^2$ which implies $4n+2 \leq (k+1)^2$. Since $4n+2$ cannot be a perfect square, so $4n+2 < (k+1)^2$, thus, $4n+3 \leq (k+1)^2$. Since $4n+3$ cannot be a perfect square also, so $4n+3 < (k+1)^2$, hence

$$k^2 < (4n+3) < (k+1)^2, \quad \text{i.e.} \quad k < \lfloor \sqrt{4n+3} \rfloor < k+1,$$

therefore $\lfloor \sqrt{4n+3} \rfloor = k = \lfloor \sqrt{4n+1} \rfloor$, as desired.

3.  By using $x = \lfloor x \rfloor + \{x\}$, the given equation becomes $\lfloor x \rfloor\{x\} + \lfloor x \rfloor - \{x\} = 10$, so $(\lfloor x \rfloor - 1)(\{x\} + 1) = 9$.

Since $\lfloor x \rfloor - 1$ is an integer, so $\{x\} + 1$ is a rational number. Let $\{x\} = m/n$ with $0 \leq m < n$, then

$$(\lfloor x \rfloor - 1)(m + n) = 9n.$$

When $m = 0$, then $\{x\} = 0$, so $x = \lfloor x \rfloor = 10$.

When $m > 0$, letting $(m, n) = 1$, then $(m + n, n) = 1$, so $(m + n) \mid 9$.

(i)   When $m + n = 3$, then $m = 1, n = 2$, so $\{x\} = \dfrac{1}{2}, \lfloor x \rfloor = 7, x = 7\dfrac{1}{2}$.

(ii)   When $m + n = 9$, then there are three possible cases:

$$m = 1, n = 8 \to \{x\} = \frac{1}{8}, \lfloor x \rfloor = 9, x = 9\frac{1}{8};$$

$$m = 2, n = 7 \to \{x\} = \frac{2}{7}, \lfloor x \rfloor = 8, x = 8\frac{2}{7};$$

$$m = 4, n = 5 \to \{x\} = \frac{4}{5}, \lfloor x \rfloor = 6, x = 6\frac{4}{5}.$$

Thus, there are five solutions for $x$: $10, 7\dfrac{1}{2}, 9\dfrac{1}{8}, 8\dfrac{2}{7}, 6\dfrac{4}{5}$.

4.  Let $10^{31} = t$, then

$$\left\lfloor \frac{10^{93}}{10^{31} + 3} \right\rfloor = \left\lfloor \frac{t^3}{t+3} \right\rfloor = \left\lfloor \frac{t^3 + 3^3}{t+3} - \frac{3^3}{t+3} \right\rfloor$$

$$= t^2 - 3t + 3^2 + \left\lfloor -\frac{3^3}{t+3} \right\rfloor = t^2 - 3t + 3^2 - 1$$

$$= t(t-3) + 8 = 10^{31}(10^{31} - 3) + 8,$$

thus, the last two digits of $\left\lfloor \dfrac{10^{93}}{10^{31} + 3} \right\rfloor$ is 08.

5. The given equation implies that $x \neq 0$ and $\lfloor x \rfloor \neq 0$ and

$$(x - \lfloor x \rfloor) + \left( \frac{92}{x} - \frac{92}{\lfloor x \rfloor} \right) = 0,$$

$$(x - \lfloor x \rfloor) \left( 1 - \frac{92}{x \lfloor x \rfloor} \right) = 0.$$

When $x = \lfloor x \rfloor$, then $x$ can be any non-zero integer.

When $\dfrac{92}{x \lfloor x \rfloor} = 1$, write $\{x\} = \alpha > 0$ and $\lfloor x \rfloor = n$, then

$$n(n + \alpha) = 92.$$

If $n > 0$, $n^2 \le 92 < n(n + 1)$ has no integer solution for $n$, if $n < 0$, then $n(n + 1) < 92 \le n^2$ has integer solution $n = -10$. Then $\alpha = 0.8$ and $x = -10 + 0.8 = -9.2$.

Thus, the solutions are $-9.2$ or any non-zero integer.

# Solutions to Testing Questions    23

## Testing Questions    (23-A)

1. $x = \dfrac{3k + 12}{k} = 3 + \dfrac{12}{k}$ implies $k \mid 12$, so $k$ may be $1, 2, 3, 4, 6, 12$, the answer is (D).

2. The given equation implies $y \equiv 1 \pmod{10}$, only (C) is possible. By checking, (C) is a solution. Thus, the answer is (C).

3. The given equality yields $\dfrac{3A + 11B}{33} = \dfrac{17}{33}$, so

$$3A + 11B = 17.$$

It is easy to find the special solution $(2, 1)$ for $(A, B)$. Since the general solution is

$$A = 2 + 11t, \quad B = 1 - 3t, \quad t \in \mathbb{Z},$$

$A \ge 1$ and $B \ge 1$ implies that $t = 0$ is the unique permitted value of $t$, so $(2, 1)$ is the unique desired solution for $(A, B)$. Thus, $A^2 + B^2 = 5$.

4.  The given conditions gives the relation $16m + 13 = 125n + 122$ for some positive integers $m$ and $n$ such that both sides are four digit numbers. Then

    $$m = \frac{125n + 109}{16} = 7n + 6 + \frac{13(n + 1)}{16}$$

    is an positive integer. The minimum value of $n$ satisfying the requirement is $n = 15$, so $m = 124$ and the four digit number is $16 \times 124 + 13 = 1997$.

5.  Let $x$ and $y$ be the numbers of dragonflies and spiders respectively. Then

    $$6x + 8y = 46 \quad \text{or} \quad 3x + 4y = 23.$$

    It is clear that $x \neq 0$ and $y \neq 0$ since 46 is neither divisible by 6 nor by 8. $4y < 23$ implies $y \leq 5$, so corresponding to $y = 1, 2, 3, 4, 5$, the equation gives

    $$x = \frac{23 - 4y}{3}$$

    can be positive integer at $y = 2$ or $y = 5$ only. $x = 5$ if $y = 2$ and $x = 1$ at $y = 5$. So the answer is that 5 dragonflies and 2 spiders, or one dragonfly and 5 spiders.

6.  The question is the same as finding the number of non-negative integer solutions for $(x, y, z)$ of the equation

    $$x + 2y + 5z = 100.$$

    It is clear that $0 \leq z \leq 20$. For any possible value of $z$, $x + 2y = 100 - 5z$. Let $u = 100 - 5z \geq 0$. Then for solving the equation $x + 2y = u$, $(-u, u)$ is a special solution. So the general solution for $(x, y)$ is

    $$x = -u + 2t, \quad y = u - t, \quad t \in \mathbb{Z}.$$

    If $u = 2k$, then $k = \frac{u}{2} \leq t \leq u = 2k$, i.e. there are $k + 1 = \frac{u}{2} + 1$ solutions for $(x, y)$.

    If $u = 2k + 1$, then $k + 1 = \frac{u+1}{2} \leq t \leq u = 2k + 1$, i.e. there are $k + 1 = \frac{u + 1}{2}$ solutions for $(x, y)$.

    Thus, we have the following table:

    $u = 100 \quad 95 \quad 90 \quad 85 \quad 80 \quad 75 \quad 70 \quad 65 \quad 60 \quad \cdots 15, \quad 10 \quad 5 \quad 0$

    $k + 1 = 51 \quad 48 \quad 46 \quad 43 \quad 41 \quad 38 \quad 36 \quad 33 \quad 31 \quad \cdots 8 \quad 6 \quad 3 \quad 1$

    The total number of solutions is given by

    $$4(10 + 20 + 30 + 40) + 5(8 + 6 + 3 + 1) + 51 = 541.$$

    Thus, there are 541 ways to get the 10 dollars.

7. Let the four digit number be $\overline{abcd} = 1000a + 100b + 10c + d$. Then

$$1000a + 100b + 10c + d + a + b + c + d = 2006,$$
$$1001a + 101b + 11c + 2d = 2006,$$

which implies that $a = 1$, so $101b + 11c + 2d = 1005$. Since

$$101 \times 8 + 11 \times 9 + 2 \times 9 = 925 < 1005,$$

therefore $b = 9$, so $11c + 2d = 96$. $11 \times 7 + 2 \times 9 = 95 < 96$ and $c < 9$ then gives $c = 8$, so $2d = 8$ i.e. $d = 4$. Thus, the four digit number is 1984.

8. From $30 < 5n + 3 < 40$ it follows that $\dfrac{27}{5} < n < \dfrac{37}{5}$ i.e. $5 < n < 8$, so $n = 6$ or 7.

When $n = 6$, then $3m = 5n + 1 = 31$ implies $m$ is not an integer, so $n = 7$.

When $n = 7$, then $3m = 5n + 1 = 36$, i.e. $m = 12$. Thus, $mn = 84$.

9. Let the numbers of roosters, hens and chicks that the buyer bought be $x, y, z$ respectively, then

$$5x + 3y + \tfrac{1}{3}z = 100,$$
$$x + y + z = 100.$$

or

$$15x + 9y + z = 300, \qquad (30.4)$$
$$x + y + z = 100. \qquad (30.5)$$

$(30.4) - (30.5)$ yields $14x + 8y = 200$ or, equivalently,

$$7x + 4y = 100. \qquad (30.6)$$

It has a special solution $x_0 = -100$, $y_0 = 200$, so the general solution for $(x, y)$ is $x = -100 + 4t$, $y = 200 - 7t$, where $t$ is any integer. The equation (30.6) indicates that $4 \mid x$, so $x$ may be $0, 4, 8, 12, 16, 20$ only. Correspondingly, $t = 25, 26, 27, 28, 29, 30$ only. Since $y \geq 0$ implies $t \leq 200/7 < 29$, so $t = 25, 26, 27, 28$ only, i.e. $y = 25, 18, 11, 4$. Thus, $x + y$ is $25, 22, 19, 16$ respectively, so $z = 75, 78, 81, 84$ correspondingly. Thus, the solutions for $(x, y, z)$ are

$$(0, 25, 75), \quad (4, 18, 78), \quad (8, 11, 81), \quad (12, 4, 84).$$

10.  $x > y > z \geq 664$ implies that $z \geq 664$, $y \geq 665$, $x \geq 666$. Since $2x + 3y + 4z = 5992$, $y$ is even, i.e. $y \geq 666$. Since $669 + 668 + 664 = 2001 > 1998$, so $y < 668$, i.e. $y = 666$, hence

$$2x + 4z = 5992 - 3 \times 666 = 3994,$$

i.e. $x + 2z = 1997$. Therefore $x$ is odd, hence, from the first equation, $z$ is also odd. $664 < z < 666$ implies $z = 665$, then $x = 667$. Thus, the answer is $x = 667$, $y = 666$, $z = 665$.

### Testing Questions    (23-B)

1.  Let the numbers of weights of 1 g, 10 g, 50 g be $x, y, z$ respectively. Then the conditions give

$$\begin{cases} x + y + z = 100, \\ x + 10y + 50z = 500. \end{cases}$$

By eliminating $x$, it follows that $9y + 49z = 400$, so

$$9(y + 5z) = 4(100 - z).$$

It implies that $4 \mid y + 5z$ and $9 \mid 100 - z$, i.e.

$$\frac{y + 5z}{4} = \frac{100 - z}{9} = t \in \mathbb{Z},$$
$$\therefore z = 100 - 9t, \quad y = 4t - 5z = 49t - 500.$$

The inequalities $y \geq 1, z \geq 1$ implies that $10 < \dfrac{501}{49} \leq t \leq 11$, so $t = 11$ and

$$z = 1, \quad y = 39, \quad x = 60.$$

Thus, in the 100 weights there are 60 of 1 g, 39 of 2 g, and 1 of 50 g, respectively.

2.  Let $(x, y, z)$ be a solution with distinct components. For letting the solution satisfy the requirement that any two components have a product which is divisible by the remaining component, let

$$x = mn, \quad y = nk, \quad z = mk,$$

where $m, n, k$ are distinct positive integers. Then given equation yields

$$n(m - k) = 1 - mk \text{ or equivalently, } n(k - m) = mk - 1.$$

Now let $k - m = 1$, then $n = mk - 1 = m(m+1) - 1 = m^2 + m - 1, k = m + 1$, so

$$x = m(m^2 + m - 1), y = (m^2 + m - 1)(m + 1), z = m(m + 1),$$

where $m$ is any natural number. The conclusion is proven.

3. We prove the conclusion by contradiction. Suppose that $(x_0, y_0)$ is a non-negative integer solution of the given equation. From

$$a(x_0 + 1) + b(y_0 + 1) = ab,$$

it follows that $b \mid a(x_0 + 1)$ and $a \mid b(y_0 + 1)$. Since $(a, b) = 1$, so $b \mid (x_0 + 1)$ and $a \mid (y_0 + 1)$. $x_0 + 1 > 0$ and $y_0 + 1 > 0$ implies that $a \leq y_0 + 1$ and $b \leq x_0 + 1$, hence

$$ab = a(x_0 + 1) + b(y_0 + 1) \geq ab + ba = 2ab,$$

a contradiction. The conclusion is proven.

4. From Theorem III, the general solution of the given equation is

$$x = x_0 + bt, \qquad y = y_0 - at, \ t \in \mathbb{Z},$$

where $(x_0, y_0)$ is a special solution. We claim that there must be a solution $(x_1, y_1)$ with $0 \leq x_1 \leq b - 1$.

The conclusion is true if $x_0 \in [0, b - 1]$. Otherwise, $x_0 \leq -1$ or $x_0 \geq b$. Since $x$ plus $b$ or minus $b$ as $t$ plus 1 or minus 1, so corresponding to some value of $t$, the value of $x$ must enter the interval $[0, b - 1]$. When $0 \leq x_1 \leq b - 1$, then

$$by_1 = c - ax_1 > ab - a - b - ax_1 \geq ab - a - b - a(b - 1) = -b,$$

so that $y_1 > -1$, i.e. $y_1 \geq 0$. Thus, $(x_1, y_1)$ is a non-negative integer solution. The conclusion is proven.

5. (i) Since every time the number obtained after operation is always odd, so it cannot be divisible by 1980.

(ii) Below we show that the number obtained after 100 times of operations may be divisible by 1981.

Let the original natural number be $x - 1$. Then after the first operation the number becomes $2(x - 1) + 1 = 2x - 1$.

The number obtained after the second operation then is $2(2x - 1) + 1 = 2^2 x - 1$.

In general, if the number obtained after the $k$th operation is $2^k x - 1$, then the number obtained after $k + 1$th operation is

$$2(2^k x - 1) + 1 = 2^{k+1} x - 1.$$

So the number obtained after $100$ times of operations is $2^{100} x - 1$. Since $(2^{100}, 1981) = 1$, so the Diophantine equation

$$2^{100} x - 1981 y = 1$$

must have integer solution $(x_0, y_0)$, and its general solution is

$$x = x_0 + 1981t, \qquad y = y_0 + 2^{100} t, \qquad t \in \mathbb{Z}.$$

Regardless of the size of $|x_0|$ and $|y_0|$, it is always possible to let $t_0$ be large enough, such that

$$x_0 + 1981 t_0 > 0 \quad \text{and} \quad y = y_0 + 2^{100} t_0 > 0.$$

Thus, it is true that $2^{100} x - 1 = 1981 y$ for this solution $(x, y)$, i.e. $1981 \mid (2^{100} x - 1)$.

## Solutions to Testing Questions    24

### Testing Questions    (24-A)

1.  In the first equation, since $(2003)^2 - 2002 \times 2004 - 1 = 0$, so $x - 1$ is a factor of the left hand side. By cross multiplication, it is obtained that

$$(x - 1)(2003^2 x + 1) = 0,$$

so the other root is $-\dfrac{1}{2003^2}$, and the larger root $m$ is $1$.

For the second equation, Since $1 + 2002 - 2003 = 0$, $x - 1$ is a factor of the left hand side, so it follows that

$$(x - 1)(x + 2003) = 0,$$

so $n$ is $-2003$. Thus $m - n = 1 + 2003 = 2004$, the answer is (A).

2.  The partition points of the range of $x$ is $-3$ and $3$.

    (i)   When $x \le -3$, then $x^2 - 2x - 24 = 0$, so $x_1 = -4$ ($x_2 = 6$ is N.A.).

   (ii) When $-3 < x \le 3$, then $x^2 - 18 = 0$, so $x = \pm 3\sqrt{2}$ (Both are N.A.).

   (iii) When $3 < x$, then $x^2 + 2x - 24 = 0$, so $x = 4$ ($x = -6$ is N.A.).

   Thus, the roots are $-4, 4$.

3.   When $m = 2$, the equation becomes $-5x - 5 = 0$, the solution is $x = -1$.

    When $m \ne 2$, $\Delta = [-(m+3)]^2 + 4(2m+1)(m-2) = 9m^2 - 6m + 1 = (3m-1)^2 \ge 0$, so

$$x_1 = \frac{(m+3) - (3m-1)}{2(m-2)} = -1, \ x_2 = \frac{(m+3) + (3m-1)}{2(m-2)} = \frac{2m+1}{m-2}.$$

4.   $a^2 - 3a + 1 = 0$ yields $a^2 + 1 = 3a$ and $a \ne 0$. Therefore by division,

$$\frac{2a^5 - 5a^4 + 2a^3 - 8a^2}{a^2 + 1} = \frac{(a^2 - 3a + 1)(2a^3 + a^2 + 3a) - 3a}{a^2 + 1}$$

$$= -\frac{3a}{3a} = -1.$$

5.   Let $x_0$ be the common root of the two given equations, then

$$1988x_0^2 + bx_0 + 8891 = 8891x_0^2 + bx_0 + 1988,$$
$$(8891 - 1988)x_0^2 = (8891 - 1988),$$
$$\therefore x_0 = \pm 1.$$

  Substituting back such $x_0$ into the first equation,

$$\pm b = -(8891 + 1988) = -10879, \ \therefore b = \mp 10879.$$

6.   (i)  When $m = 1$, then $-6x + 1 = 0$, so there is a real root $x = \frac{1}{6}$.

    (ii)  When $m = -1$, then $-2x + 1 = 0$, so there is a real root $x = \frac{1}{2}$.

    (iii)  When $m^2 - 1 \ne 1$, then $\Delta = 4(m+2)^2 - 4(m^2 - 1) = 16m + 20 \ge 0$ implies that $m \ge -\frac{5}{4}$ and $m \ne \pm 1$.

  Thus, the range of $m$ is $m \ge -\frac{5}{4}$.

7.   Let $x_0$ be the common root, then $x_0^2 - kx_0 - 7 = 0$ and $x_0^2 - 6x_0 - (k+1) = 0$. Then their difference gives

$$(6 - k)x_0 - (6 - k) = 0, \quad \text{or} \quad (6 - k)(x_0 - 1) = 0.$$

Notice that $k \neq 6$. Otherwise, the two equations are identical so that they have two common roots. Therefore $x_0 = 1$, and it implies that $k = -6$. By substituting back the value of $k$ into the given equations, it follows that

$$x^2 + 6x - 7 = 0 \quad \text{and} \quad x^2 - 6x + 5 = 0,$$
$$\therefore (x - 1)(x + 7) = 0 \quad \text{and} \quad (x - 1)(x - 5) = 0.$$

Thus, the different roots are $-7$ and $5$ respectively.

8.  Since the given equation has two equal real roots implies that its discriminant is 0, so

$$\Delta = (2b)^2 - 4(c + a)(c - a) = 4(b^2 + a^2 - c^2) = 0,$$
$$\therefore a^2 + b^2 = c^2.$$

From $c^2 = a^2 + b^2 < a^2 + b^2 + 2ab = (a + b)^2$, it follows that $c < a + b$, so the three segments with lengths $a, b, c$ can form a triangle. Further, from Pythagoras' Theorem, the triangle is a right-angled triangle with the hypotenuse side of length $c$.

9.  Since the discriminant of the equation is non-negative,

$$4(1 + a)^2 - 4(3a^2 + 4ab + 4b^2 + 2) \geq 0,$$
$$(1 + a)^2 - (3a^2 + 4ab + 4b^2 + 2) \geq 0,$$
$$2a^2 - 2a + 1 + 4ab + 4b^2 \leq 0,$$
$$(a - 1)^2 + (a + 2b)^2 \leq 0,$$
$$\therefore a = 1, \quad \text{and} \quad a + 2b = 0,$$
$$a = 1, \quad b = -\frac{1}{2}.$$

10. The discriminant of the equation is given by

$$\begin{aligned}
\Delta &= (a + b + c)^2 - 4(a^2 + b^2 + c^2) \\
&= -3a^2 - 3b^2 - 3c^2 + 2ab + 2bc + 2ca \\
&= -(a^2 - 2ab + b^2) - (b^2 - 2bc + c^2) \\
&\quad -(c^2 - 2ca + a^2) - (a^2 + b^2 + c^2) \\
&= -[(a - b)^2 + (b - c)^2 + (c - a)^2 + (a^2 + b^2 + c^2)] < 0.
\end{aligned}$$

so the equation has no real roots, the answer is (D).

## Testing Questions (24-B)

1. Mr. Fat has the winning strategy. A set of three distinct rational nonzero numbers $a, b$, and $c$, such that $a + b + c = 0$, will do the trick. Let $A, B$, and $C$ be any arrangement of $a, b$, and $c$, and let $f(x) = Ax^2 + Bx + C$. Then

$$f(1) = A + B + C = a + b + c = 0,$$

which implies that 1 is a solution.

Since the product of the two solutions is $\dfrac{C}{A}$, the other solution is $\dfrac{C}{A}$, and it is different from 1.

2. By $\Delta_i, i = 1, 2, 3$ we denote the discriminant of the $i$th equation. It suffices to show $\Delta_1 + \Delta_2 + \Delta_3 > 0$. Then

$$
\begin{aligned}
\Delta_1 + \Delta_2 + \Delta_3 &= (4b^2 - 4ac) + (4c^2 - 4ab) + (4a^2 - 4bc) \\
&= 2(2a^2 + 2b^2 + 2c^2 - 2ab - 2bc - 2ca) \\
&= 2[(a - b)^2 + (b - c)^2 + (c - a)^2] > 0.
\end{aligned}
$$

Thus, the conclusion is proven.

3. From the assumptions $a > 1$ and $b > 1$ and $a \neq b$. Let $x_0$ be the common root, then

$$(a-1)x_0^2 - (a^2+2)x_0 + (a^2+2a) = 0, \ (b-1)x_0^2 - (b^2+2)x_0 + (b^2+2b) = 0.$$

Notice that $x_0 \neq 1$. Otherwise, if $x_0 = 1$, then above two equalities becomes $a = 1 = b$.

After eliminating the term of $x_0^2$ and doing simplification, it follows that

$$(a - b)(ab - a - b - 2)(x_0 - 1) = 0.$$

Since $a - b \neq 0$ and $x_0 \neq 1$, so $ab - a - b - 2 = 0$ i.e. $ab = a + b + 2$.

(i) When $a > b > 1$, then $b = 1 + \dfrac{b}{a} + \dfrac{2}{a} < 3$, so $b = 2, a = \dfrac{4}{b - 1} = 4$.

(ii) When $b > a > 1$, then, by symmetry, $a = 2, b = 4$.

Thus, in each of above two cases,

$$\frac{a^b + b^a}{a^{-b} + b^{-a}} = (a^b + b^a) \cdot \frac{a^b b^a}{a^b + b^a} = a^b b^a = 256.$$

4.   Since the left hand side cannot be zero, so $m \neq 0$ and $x \neq 0$.

   (i)   When $m > 0$, then $x > 0$.

   For $0 < x \le 1$, then $1 - x^2 + 4 - x^2 = mx$, i.e. $2x^2 + mx - 5 = 0$, so

   $$x = \frac{-m + \sqrt{m^2 + 40}}{4} \le 1.$$

   It implies that $m^2 + 40 \le m^2 + 8m + 16$, i.e. $3 \le m$.

   For $1 < x \le 2$, then $x^2 - 1 + 4 - x^2 = mx$, so $1 < x = \dfrac{3}{m} \le 2$, i.e.

   $$\frac{3}{2} \le m < 3.$$

   for $2 < x$, then $2x^2 - mx - 5 = 0$, so $x = \dfrac{m + \sqrt{m^2 + 40}}{4} > 2$, it implies
   that

   $$m^2 + 40 > m^2 - 16m + 64, \quad \text{i.e. } m > \frac{3}{2}.$$

   (ii)   When $m < 0$, then $x < 0$, so the case can be converted to the case (i)
   if use $-m, -x$ to replace $m, x$ respectively. Thus, the solutions are

   $$x = \begin{cases} \dfrac{3}{m} \text{ or } \dfrac{m + \sqrt{m^2 + 40}}{4} & \text{if } \dfrac{3}{2} \le m < 3, \\[2mm] \dfrac{-m + \sqrt{m^2 + 40}}{4} & \text{if } m \ge 3, \\[2mm] \dfrac{3}{m} \text{ or } \dfrac{m - \sqrt{m^2 + 40}}{4} & \text{if } -3 < m \le -\dfrac{3}{2}, \\[2mm] \dfrac{-m + \sqrt{m^2 + 40}}{4} & \text{if } m \le -3, \end{cases}$$

   and has no solution for other values of $m$.

5.   Suppose that the first equation has no two distinct real roots, then

   $$\Delta_1 = 1 - 4q_1 \le 0, \quad \text{i.e. } q_1 \ge \frac{1}{4}.$$

   In this case, then the discriminant of the second equation is

   $$\begin{aligned} \Delta_2 &= p^2 - 4q_2 = (q_1 + q_2 + 1)^2 - 4q_2 \\ &= q_2^2 + 2q_2(q_1 + 1) + (q_1 + 1)^2 - 4q_2 \\ &= q_2^2 + 2q_2(q_1 - 1) + (1 + q_1)^2. \end{aligned}$$

To show $\Delta_2$ is always positive, consider the last expression as a quadratic function of $q_2$, then its discriminant is

$$\Delta_3 = 4(a_1 - 1)^2 - 4(1 + q_1)^2 = -16q_1 \leq -4,$$

so $\Delta_2 > 0$ for any value of $q_2$, i.e. the second equation has two distinct real roots, the conclusion is proven.

## Solutions to Testing Questions    25

### Testing Questions    (25-A)

1. Let $x_1 = 2t$, $x_2 = 3t$, where $t$ is some real number. By Viete theorem,

$$5t = x_1 + x_2 = \frac{5}{2} \Rightarrow t = \frac{1}{2},$$

$$\therefore x_2 - x_1 = t = \frac{1}{2}.$$

2. Let

$$x^2 + px + q = 0, \tag{30.7}$$

$$x^2 + 2qx + \frac{1}{2}p = 0, \tag{30.8}$$

and let $\alpha, \beta$ be the roots of (30.7), then the roots of (30.8) are $\alpha - 1, \beta - 1$. By Viete Theorem,

$$\alpha + \beta = -p, \tag{30.9}$$

$$\alpha\beta = q, \tag{30.10}$$

$$\alpha + \beta - 2 = -2q, \tag{30.11}$$

$$(\alpha - 1)(\beta - 1) = \frac{1}{2}p. \tag{30.12}$$

By (30.9) − (30.11),

$$-p + 2q = 2. \tag{30.13}$$

By (30.10) − (30.11) − (30.12),

$$-p + 6q = 2. \tag{30.14}$$

Then (30.14) − (30.13) yields $q = 0$, $p = -2$, hence the equation (30.7) is $x^2 - 2x = 0$, its roots are $\alpha = 0, \beta = 2$. Similarly, the equation (30.8) is $x^2 - 1 = 0$, its roots are $x_1 = -1 = \alpha - 1$, $x_2 = 1 = \beta - 1$.

3.  Viete theorem produces

$$\alpha + \beta = p, \quad \alpha\beta = q, \quad \alpha^2 + \beta^2 = p, \quad \alpha^2\beta^2 = q.$$

From $q^2 = q$, it follows that $q = 0$ or $1$. Since $p = (\alpha + \beta)^2 - 2\alpha\beta = p^2 - 2q$,

When $q = 0$, then $p = 0$ or $1$. So $(p, q) = (0, 0)$ or $(1, 0)$.

When $q = 1$, then $p = -1$ or $2$. But $x^2 + x + 1 = 0$ has no real root when $p = -1$, so $p = 2$ and $(p, q) = (2, 1)$.

Thus, there are three desired pairs for $(p, q)$.

4.  Let $(x_1, x_2)$ be an integer solution of the given solution. Then $x_1 + x_2 = -p, x_1 x_2 = q$, so

$$198 = p + q = -(x_1 + x_2) + x_1 x_2 = (x_1 - 1)(x_2 - 1) - 1,$$
$$\therefore (x_1 - 1)(x_2 - 1) = 199 = 1 \cdot 199 = (-1) \times (-199).$$

Letting $x_1 \le x_2$, then $x_1 - 1 = 1, x_2 - 1 = 199$ or $x_1 - 1 = -199, x_2 - 1 = -1$, so the solutions are

$$(x_1, x_2) = (2, 200) \quad \text{or} \quad (-198, 0).$$

5.  Let $\alpha, \beta$ be roots of the equation, then $\alpha + \beta = -\dfrac{a}{2}, \alpha\beta = \dfrac{-2a + 1}{2}$, and $\alpha^2 + \beta^2 = 7\dfrac{1}{4}$, so

$$(\alpha + \beta)^2 = 7\frac{1}{4} + 2(\frac{-2a + 1}{2}) = \frac{33}{4} - 2a,$$

$$\frac{a^2}{4} = \frac{33}{4} - 2a,$$

$$a^2 + 8a - 33 = 0 \Rightarrow (a - 3)(a + 11) = 0 \Rightarrow a = 3 \text{ or } -11.$$

Since $\Delta = a^2 + 16a - 8 \ge 0$, so

$$a \le \frac{-16 - \sqrt{288}}{2} < -11, \text{ or } a \ge \frac{-16 + \sqrt{288}}{2} > 0,$$

thus $a = 3$.

6.  $\alpha^2 = 2\alpha + 1$ yields $\alpha^4 = 4\alpha^2 + 4\alpha + 1 = 12\alpha + 5$, and similarly $\beta^2 = 2\beta + 1, \beta^3 = 2\beta^2 + \beta = 5\beta + 2$. Viete Theorem gives $\alpha + \beta = 2$ and $\alpha\beta = 1$, thus,

$$\begin{aligned} 5\alpha^4 + 12\beta^3 &= 5(12\alpha + 5) + 12(5\beta + 2) = 60(\alpha + \beta) + 49 \\ &= 120 + 49 = 169. \end{aligned}$$

7. Viete Theorem gives $\alpha + \beta = -19, \alpha\beta = -97$. Then

$$-\frac{m}{n} = \frac{1+\alpha}{1-\alpha} + \frac{1+\beta}{1-\beta} = \frac{(1+\alpha)(1-\beta) + (1-\alpha)(1+\beta)}{(1-\alpha)(1-\beta)}$$

$$= \frac{2(1-\alpha\beta)}{1-(\alpha+\beta)+\alpha\beta} = \frac{2 \cdot 98}{77} = -\frac{28}{11},$$

so $m = 28, n = 11$ and $m + n = 39$.

8. First of all, $\alpha + \beta = -(a+b)$ and $\alpha\beta = \frac{4}{3}ab$. The condition $\alpha(\alpha+1) + \beta(\beta+1) = (\alpha+1)(\beta+1)$ implies

$$\alpha^2 + \beta^2 - \alpha\beta = 1 \quad \text{i.e.} \quad (\alpha+\beta)^2 - 3\alpha\beta = 1,$$
$$\therefore (a+b)^2 - 4ab = 1, \quad \text{i.e.} \quad (a-b)^2 = 1, \quad \text{so}$$
$$a - b = 1.$$

$\Delta \geq 0$ implies $3(a+b)^2 \geq 16ab = 4[(a+b)^2 - 1]$, so $(a+b)^2 \leq 4$, i.e.

$$-2 \leq a+b \leq 2.$$

Since $a - b = 1$, so $-1 \leq 2a \leq 3$, i.e. $a = 0$ or $1$. Hence,

$$(a,b) = (0,-1) \quad \text{or} \quad (1,0).$$

By checking, the two solutions satisfy the requirements. Thus, there are two desired pairs for $(a, b)$.

9. Changing the second equality in the form of $19\left(\frac{1}{t}\right)^2 + 99\left(\frac{1}{t}\right) + 1 = 0$, it follows that $s$ and $1/t$ both are roots of the equation $19x^2 + 99x + 1 = 0$. Therefore, by Viete Theorem,

$$s + \frac{1}{t} = -\frac{99}{19}, \quad \text{and} \quad \frac{s}{t} = \frac{1}{19}.$$
$$\therefore \frac{st + 4s + 1}{t} = s + \frac{1}{t} + 4\frac{s}{t} = -\frac{99}{19} + \frac{4}{19} = -5.$$

10. Viete Theorem gives $\alpha + \beta = -p, \alpha\beta = 1, \gamma + \delta = -q, \gamma\delta = 1$. So

$$(\alpha - \gamma)(\beta - \gamma)(\alpha + \delta)(\beta + \delta) = [(\alpha - \gamma)(\beta + \delta)][(\beta - \gamma)(\alpha + \delta)]$$
$$= (\alpha\beta + \alpha\delta - \beta\gamma - \gamma\delta)(\alpha\beta + \beta\delta - \alpha\gamma - \gamma\delta)$$
$$= (\alpha\delta - \beta\gamma)(\beta\delta - \alpha\gamma) = \alpha\beta\delta^2 - \alpha^2\gamma\delta - \beta^2\gamma\delta + \alpha\beta\gamma^2$$
$$= \delta^2 - \alpha^2 - \beta^2 + \gamma^2 = [(\delta + \gamma)^2 - 2\delta\gamma] - [(\alpha + \beta)^2 - 2\alpha\beta]$$
$$= (q^2 - 2) - (p^2 - 2) = q^2 - p^2.$$

### Testing Questions   (25-B)

1.  Let $A = \dfrac{2}{\alpha} + 3\beta^2$, $B = \dfrac{2}{\beta} + 3\alpha^2$. Since $\alpha + \beta = 7$, $\alpha\beta = 8$, and $\alpha - \beta = \sqrt{\Delta} = \sqrt{49 - 32} = \sqrt{17}$, it follows that

$$A + B = \frac{2(\alpha + \beta)}{\alpha\beta} + 3(\alpha^2 + \beta^2) = \frac{14}{8} + 3(7^2 - 16) = \frac{403}{4},$$

$$A - B = -\frac{2(\alpha - \beta)}{\alpha\beta} - 3(\alpha^2 - \beta^2) = -(\alpha - \beta)\left[\frac{2}{\alpha\beta} + 3(\alpha + \beta)\right]$$

$$= -\sqrt{\Delta}\left[\frac{1}{4} + 21\right] = -\frac{85\sqrt{17}}{4},$$

therefore

$$\frac{2}{\alpha} + 3\beta^2 = A = \frac{1}{2}\left[\frac{403}{4} - \frac{85\sqrt{17}}{4}\right] = \frac{403 - 85\sqrt{17}}{8}.$$

2.  $a = 8 - b$ and $c^2 = ab - 16$ yields $a + b = 8$, $ab = c^2 + 16$. Then, by inverse Viette Theorem, $a, b$ are the real roots of the equation

$$x^2 - 8x + (c^2 + 16) = 0.$$

Since its discriminant $\Delta$ is non-negative, i.e. $(-8)^2 - 4(c^2 + 16) \geq 0$, then

$$8^2 \geq 4(c^2 + 16),$$
$$4c^2 \leq 0, \quad \therefore c = 0,$$

therefore $a, b$ are the roots of the equation $x^2 - 8x + 16 = (x - 4)^2 = 0$, i.e. $a = b = 4$.

3.  The assumption in question implies that

$$(x - \alpha)(x - \beta) = x^2 + px + q \quad \text{and} \quad (x - \gamma)(x - \delta) = x^2 + Px + Q.$$

Consequently,

$$(\alpha - \gamma)(\beta - \gamma)(\alpha - \delta)(\beta - \delta) = [(\gamma - \alpha)(\gamma - \beta)][(\delta - \alpha)(\delta - \beta)]$$
$$= (\gamma^2 + p\gamma + q)(\delta^2 + p\delta + q).$$

However,

$$\gamma + \delta = -P, \qquad \gamma\delta = Q,$$

hence

$$(\alpha - \gamma)(\beta - \gamma)(\alpha - \delta)(\beta - \delta) = (\gamma^2 + p\gamma + q)(\delta^2 + p\delta + q)$$
$$= \gamma^2\delta^2 + p\gamma^2\delta + q\gamma^2 + p\gamma\delta^2 + p^2\gamma\delta + pq\gamma + q\delta^2 + pq\delta + q^2$$
$$= (\gamma\delta)^2 + p\gamma\delta(\gamma + \delta) + q[(\gamma + \delta)^2 - 2\gamma\delta] + p^2\gamma\delta + pq(\gamma + \delta) + q^2$$
$$= Q^2 - pPQ + q(P^2 - 2Q) + p^2Q - pqP + q^2$$
$$= Q^2 + q^2 - pP(Q + q) + qP^2 + p^2Q - 2qQ.$$

4.  Let $\alpha$ and $\beta$ be the roots of the given equation. Then

$$\alpha + \beta = -a, \quad \text{and} \quad \alpha\beta = b + 1.$$

Consequently,

$$a^2 + b^2 = (\alpha + \beta)^2 + (\alpha\beta - 1)^2 = \alpha^2 + \beta^2 + \alpha^2\beta^2 + 1$$
$$= (\alpha^2 + 1)(\beta^2 + 1),$$

which is a composite number. The conclusion is proven.

5.  This problem involves an equation of high degree. By using substitutions, it can be reduced to a quadratic equation. Let $y = \dfrac{13 - x}{x + 1}$, then the given equation becomes

$$xy(x + y) = 42.$$

The technique of solving the problem is to solve $xy$ and $x + y$ first by applying the inverse Viete Theorem. For this the value of $(xy) + (x + y)$ is needed. Since

$$xy + (x + y) = \frac{13x - x^2}{x + 1} + \frac{x^2 + x + 13 - x}{x + 1} = \frac{13x + 13}{x + 1} = 13,$$

by using inverse Viete Theorem, $xy$ and $x + y$ are the roots of the equation

$$z^2 - 13z + 42 = 0.$$

Since $z = 6$ or $7$, by solving the systems $xy = 6, x + y = 7$ and $xy = 7, x + y = 6$ respectively, the solutions for $x$ are

$$x_1 = 1, \quad x_2 = 6, \quad x_3 = 3 + \sqrt{2}, x_4 = 3 - \sqrt{2}.$$

By checking, they are all the solution of the original equation.

# Solutions to Testing Questions   26

## Testing Questions   (26-A)

1.  $a^2 + a - 6 = (a-2)(a+3)$ implies that $(a-2)$ and $(a+3)$ are both factors of 1260, and their difference is 5. Since $1260 = 2^2 \times 3^2 \times 5 \times 7$, where the pairs of two factors with the difference 5 are $(1,6), (2,7), (4,9), (7,12)$ and $(9,14)$. Thus,

$$a - 2 = 1, 2, 4, 7, 9, \quad \text{i.e.} \quad a = 3, 4, 6, 9, 11.$$

2.  The original equation yields $(x-y)(x+y) = 12$. Since $x-y$ and $x+y$ have the same parity, and $12 = 2 \times 6 = (-2) \times (-6)$, so there are four systems of simultaneous equations:

$$\begin{cases} x-y=2, \\ x+y=6, \end{cases} \quad \begin{cases} x-y=6, \\ x+y=2, \end{cases} \quad \begin{cases} x-y=-2, \\ x+y=-6, \end{cases} \quad \begin{cases} x-y=-6, \\ x+y=-2, \end{cases}$$

from which four solutions are obtained, i.e. $(4,2); (4,-2); (-4,-2);$ and $(-4,2)$. Thus, there are four required pairs.

3.  The given equations give $(x-y)(x+y) = 3^4$ and $(z-w)(z+w) = 3^4$. Since $x - y < x + y$ and $z - w < z + w$, if two of the four numbers $x-y, x+y, z-w, z+w$ have equal values, then the other two must be equal also, and it must be teh case that $x - y = z - w$ and $x + y = z + w$, but then it implies that $y - w = x - z = w - y$, i.e. $y = w$, a contradiction. Thus, $x-y, x+y, z-w, z+w$ must be four distinct values, and further, they are the four distinct factors of $3^4$ with $x+y, z+w$ being the larger two factors and $x-y, z-w$ being the smaller two. Since $3^4 = 3^4 \times 1 = 3^3 \times 3$, so

$$xz + yw + xw + yz = (x+y)(z+w) = 3^4 \cdot 3^3 = 3^7 = 2187.$$

4.  By eliminating the denominators, the given equation becomes

$$15 + 3x - 2xy = 2x^2 y,$$
$$2x^2 y - 3x + 2xy - 15 = 0,$$
$$(2xy - 3)(x + 1) = 12 = 1 \cdot 12 = -1 \times -12 = 3 \cdot 4 = (-3) \times (-4)$$

since $2xy - 3$ is odd, hence there are four possible systems:

$$\begin{cases} 2xy - 3 &= 1, \\ x + 1 &= 12, \\ 2xy - 3 &= 3, \\ x + 1 &= 4, \end{cases} \quad \begin{cases} 2xy - 3 &= -1, \\ x + 1 &= -12, \\ 2xy - 3 &= -3, \\ x + 1 &= -4, \end{cases}$$

The first, second, and the fourth systems have no integer solutions, the third system has the solution $x = 3, y = 1$. Thus, there is exactly one integer solution $x = 3, y = 1$.

5.  Simplify the equation to the form $xy + 42 = 9y$, then

$$y = \frac{42}{9-x}.$$

$y$ is a positive integer implies that $9 - x$ is a positive divisor of 42, so

$$9 - x = 1, 2, 3, 6, 7, \quad \text{i.e. } x = 8, 7, 6, 3, 2.$$

Correspondingly, $y = 42, 21, 14, 7, 6$. By checking, the five solutions satisfy the original equation, so the answer is 5.

6.  The given equation yields $8y - 12x = xy$, so $xy + 12x - 8y - 96 = -96$, i.e.

$$(x - 8)(y + 12) = -96.$$

Since $y + 12 \geq 13$ and $-7 \leq x - 8 < 0$, there are five possible cases to be considered:

$$\begin{cases} x - 8 = -1, \\ y + 12 = 96, \end{cases} \quad \begin{cases} x - 8 = -2, \\ y + 12 = 48, \end{cases} \quad \begin{cases} x - 8 = -3, \\ y + 12 = 32, \end{cases}$$
$$\begin{cases} x - 8 = -4, \\ y + 12 = 24, \end{cases} \quad \begin{cases} x - 8 = -6, \\ y + 12 = 16, \end{cases}$$

From them five positive integer solutions are obtained easily:

$$(7, 84), \quad (6, 36), \quad (5, 20), \quad (4, 12), \quad (2, 4).$$

Thus, the answer is 5.

7.  By rewriting the equation in the form $x^2 - (2 + y)x + y^2 - 2y = 0$, and considering it as a quadratic equation in $x$ ($y$ is considered as a constant in its range), then the equation has integer solutions in $x$, so its discriminant is a perfect square.

$$\Delta = (2 + y)^2 - 4(y^2 - 2y) = 4 + 12y - 3y^2 = 16 - 3(y - 2)^2 = n^2$$

for some integer $n$, it follows that $(y - 2)^2 \leq \dfrac{16}{3}$, so

$$-3 < -\frac{4}{3}\sqrt{3} \leq y - 2 < \frac{4}{3}\sqrt{3} < 3.$$

Thus, $y$ may be $0, 1, 2, 3, 4$.

When $y = 0$, then $x^2 - 2x = 0$, so $x = 0$ or 2.

When $y = 1$ or 3, then $\Delta = 13$ which is not a perfect square.

When $y = 2$, then $x^2 - 4x = 0$, so $x = 0$ or 4.

When $y = 4$, then $x^2 - 6x + 8 = 0$, so $x = 2$ or 4.

Thus, there are a total of 6 desired pairs.

8.  Suppose that the two integral roots of the given equation are $m$ and $n$. Then, by Viete Theorem,

$$m + n = p, \tag{30.15}$$
$$mn = -580p, \tag{30.16}$$

therefore one of $m$ and $n$ is divisible by $p$. Without loss of generality, we assume that $p \mid m$. Then $m = kp$ for some integer $k$. (30.15) yields $n = (1 - k)p$. then (30.16) yields $(k - 1)kp^2 = 580p$, i.e. $(k - 1)kp = 580 = 4 \times 5 \times 29$. Thus, $p = 29$.

9.  Since the sum of the squares is to be an even number, it may be reasoned that either all three of the numbers $x^2, y^2, z^2$ (hence also $x$, $y$, and $z$) are even, or one of them is even and two are odd. But in the last event, the sum would be divisible only by 2 and the product $2xyz$ would be divisible by 4. Hence we must conclude that $x$, $y$, and $z$ must all be even: $x = 2x_1, y = 2y_1, z = 2z_1$. If we substitute these into the given equation and divide through by 4, we obtain

$$x_1^2 + y_1^2 + z_1^2 = 4x_1y_1z_1.$$

As above, this equation implies that $x_1, y_1$, and $z_1$ are all even numbers, and so we can write $x_1 = 2x_2, y_1 = 2y_2, z_1 = 2z_2$, which yields the equation

$$x_2^2 + y_2^2 + z_2^2 = 2^3 x_2 y_2 z_2,$$

which, in turn, implies that $x_2, y_2$, and $z_2$ are all even numbers also.

Continuation of this process leads to the conclusion that the following set of numbers are all even:

$$x, y, z;$$

$$x_1 = \frac{x}{2}, \quad y_1 = \frac{y}{2}, \quad z_1 = \frac{z}{2};$$
$$x_2 = \frac{x}{4}, \quad y_2 = \frac{y}{4}, \quad z_2 = \frac{z}{4};$$

$$x_3 = \frac{x}{8}, \quad y_3 = \frac{y}{8}, \quad z_3 = \frac{z}{8};$$

$$\vdots$$

$$x_k = \frac{x}{2^k}, \quad y_k = \frac{y}{2^k}, \quad z_k = \frac{z}{2^k};$$

$$\vdots$$

(the numbers $x_k, y_k, z_k$ satisfy the equation $x_k^2 + y_k^2 + z_k^2 = 2^{k+1} x_k y_k z_k$). But this is possible only if $x = y = z = 0$.

10. The second equation leads to $y = 1, x + z = 31$ at once. By substituting them into the first equation, it follows that

$$x(1 + 31 - x) = 255,$$
$$x^2 - 32x + 255 = 0,$$
$$(x - 15)(x - 17) = 0,$$
$$\therefore x_1 = 15, x_2 = 17.$$

Thus the solutions are $x = 15, y = 1, z = 16$ and $x = 17, y = 1, z = 14$. The answer is (B).

## Testing Questions    (26-B)

1. By factorization, the given equation can be factorized as follows:

$$4x^3 + 4x^2 y - 15xy^2 - 18y^3 - 12x^2 + 6xy + 36y^2 + 5x - 10y$$
$$= (4x^3 - 8x^2 y) + (12x^2 y - 24xy^2) + (9xy^2 - 18y^3) - (12x^2 - 24xy)$$
$$\quad -(18xy - 36y^2) + (5x - 10y)$$
$$= (x - 2y)(4x^2 + 12xy + 9y^2 - 12x - 18y + 5)$$
$$= (x - 2y)[(2x + 3y)^2 - 6(2x + 3y) + 5]$$
$$= (x - 2y)(2x + 3y - 1)(2x + 3y - 5),$$

so

$$(x - 2y)(2x + 3y - 1)(2x + 3y - 5) = 0.$$

$x - 2y = 0$ yields the positive integer solutions $(k, 2k), k \in \mathbb{N}$.

$2x + 3y - 1 = 0$ has no positive integer solution.

$2x + 3y - 5 = 0$ has only one positive integer solution $(1, 1)$.

Thus, the solution set is $\{(1, 1)\} \cup \{(k, 2k) : \forall k \in \mathbb{N}\}$.

2.   It is easy to convert the given equation to the form

$$(x - z)(y - z) = z^2. \tag{30.17}$$

Let $t$ represent the greatest common divisor of the integers $x, y$, and $z$; that is, $x = x_1 t, y = y_1 t, z = z_1 t$, where $x_1, y_1$, and $z_1$ are a relatively prime set. Further, let $m = (x_1, z_1)$ and $n = (y_1, z_1)$. That is, we write $x_1 = mx_2, z_1 = mz_2; y_1 = ny_2, z_1 = nz_2$, where $x_2$ and $z_2$, $y_2$ and $z_2$ are relatively prime. The integers $m$ and $n$ are relatively prime, since $x_1, y_1$, and $z_1$ have no common divisor. Since $z_1$ is divisible both by $m$ and by $n$, we may write $z_1 = mnp$.

If we now substitute $x = mx_2 t, y = ny_2 t, z = mnpt$ into the basic equation (30.17), and divide the equality by $mnt^2$, it follows that

$$(x_2 - np)(y_2 - mp) = mnp^2. \tag{30.18}$$

Notice that $x_2$ is relatively prime to $p$, since $m$ is the greatest common divisor of the numbers $x_1 = mx_2$ and $z_1 = mnp$; similarly, $y_2$, and $p$ are relatively prime. Upon expanding the left hand side of (30.18), we see that $x_2 y_2 = x_2 mp + y_2 np$ is divisible by $p$. It follows that $p = 1$, and the equation takes on the form

$$(x_2 - n)(y_2 - m) = mn.$$

Now $x_2$ is relatively prime to $n$, for the three integers $x_1 = mx_2, y_1 = ny_2$, and $z_1 = mn$ are relatively prime. Consequently, $x_2 - n$ is relatively prime to $n$, whence $y_2 - m$ is divisible by $n$. Similarly, $x_2 - n$ is divisible by $m$. Thus, $x_2 - n = \pm m, y_2 - m = \pm n; x_2 = \pm y_2 = \pm m + n$. Therefore,

$$x = m(m + n)t, \qquad y = \pm n(m + n)t, \qquad z = mnt,$$

where $m, n, t$ are arbitrary integers, that is the formula for general solution.

3.   The discriminant of the quadratic equation must be a perfect square number implies that

$$(5p)^2 - 20(66p - 1) = n^2$$

for some integer $n \geq 0$, so $25p^2 - 1320p + 20 = n^2$, or $(5p - 132)^2 - n^2 = 17404 = 2^2 \cdot 19 \cdot 229$. Note that $5 \mid n$, and the equation gives $p - 1 \equiv 0 \pmod 5$, so $p = 5k + 1$ for some positive integer $k$, then

$$(25k - 127 - n)(25k - 127 + n) = 38 \cdot 458 = (-38)(-458).$$

The system $25k - 127 - n = 38, 25 - 127 + n = 458$ yields the solution

$$n = \frac{1}{2}(458 - 38) = 210, \quad k = \frac{1}{25}(38 + 127 + 210) = 15, \quad \therefore p = 76.$$

However, the system $25k - 127 - n = -458, 25k - 127 + n = -38$ has no integer solution, since from them we have $50k - 254 = -496$, so there is no integer solution for $k$.

By checking, $p = 76$ yields equation $5x^2 - 380x + 5015 = 0$ or $5(x - 17)(x - 59) = 0$, its two roots are 17 and 59. Thus, $p = 76$ is the answer.

**Note:** The problem can be solved by use of Viete Theorem

4. Let $x_1, x_2$ be the integer roots of the given equation. Viete Theorem yields

$$x_1 + x_2 = pq \text{ and } x_1 x_2 = p + q,$$

so $x_1, x_2$ are both positive integers. Then their difference yields

$$x_1 x_2 - x_1 - x_2 = p + q - pq,$$

i.e.

$$(x_1 - 1)(x_2 - 1) + (p - 1)(q - 1) = 2. \tag{30.19}$$

When the first term on the left hand side of (30.19) is 0 but the second term is 2, then

$$p - 1 = 1, q - 1 = 2 \text{ or } p - 1 = 2, q - 1 = 1,$$

so $(p, q) = (2, 3)$ or $(3, 2)$, and the equation is $x^2 - 6x + 5 = 0$, which has two integer solutions 5 and 1.

When the first term and the second term on the left hand side of (30.19) are both 1, then

$$p - 1 = 1, q - 1 = 1 \text{ i.e. } (p, q) = (2, 2),$$

so the equation is $x^2 - 4x + 4 = 0$, which has two equal integer solutions $x_1 = x_2 = 2$.

When the first term on the left hand side of (30.19) is 2 and the second term is 0, then

$$p - 1 = 0 \text{ or } q - 1 = 0$$

and $(x_1, x_2) = (2, 3)$ or $(3, 2)$, so the equation is $x^2 - 5x + 6 = 0$, which implies that $(p, q) = (1, 6)$ or $(6, 1)$. Thus,

$$(p, q) = (2, 3), \quad (3, 2), \quad (2, 2), \quad (1, 6), \quad (6, 1).$$

5. (i) When $n = m$, then $\dfrac{n^3 + 1}{n^2 - 1}$ is an integer. From

$$\frac{n^3 + 1}{mn - 1} = \frac{n^3 + 1}{n^2 - 1} = \frac{n^2 - n + 1}{n - 1} = n + \frac{1}{n - 1},$$

therefore $n = 2$, i.e. $n = m = 2$ is a solution.

(ii)    When $n \neq m$, since $m^3$ and $mn - 1$ are relatively prime and that

$$\frac{m^3(n^3 + 1)}{mn - 1} = \frac{(mn)^3 - 1}{mn - 1} + \frac{m^3 + 1}{mn - 1}$$
$$= (mn)^2 + mn + 1 + \frac{m^3 + 1}{mn - 1},$$

so $\dfrac{n^3 + 1}{mn - 1}$ and $\dfrac{m^3 + 1}{mn - 1}$ are either both integers or both non-integers. Therefore, it suffices to discuss the case $m > n$ below.

For $n = 1$, $\dfrac{n^3 + 1}{mn - 1} = \dfrac{2}{m - 1}$ is an integer, therefore $m = 2$ or 3;

For $n \geq 2$, letting $\dfrac{n^3 + 1}{mn - 1} = k$, then

$$n^3 + 1 = k(mn - 1),$$
$$1 \equiv k \cdot (-1) \pmod{n},$$
$$k \equiv -1 \pmod{n},$$

i.e., there exists a positive integer $p$ such that $\dfrac{n^3 + 1}{mn - 1} = pn - 1$.
Thus,

$$pn - 1 < \frac{n^3 + 1}{n^2 - 1} = \frac{n^2 - n + 1}{n - 1} = n + \frac{1}{n - 1},$$
$$(p - 1)n < 1 + \frac{1}{n - 1},$$
$$\therefore \quad p = 1,$$
$$\frac{n^3 + 1}{mn - 1} = n - 1,$$
$$m = \frac{n^2 + 1}{n - 1} = n + 1 + \frac{2}{n - 1},$$

i.e., $n = 2$ or 3, and hence, $m = 5$ for both cases.

Thus, the solutions $(m, n)$ are the pairs:

$$(2, 2), \quad (2, 1), \quad (3, 1), \quad (5, 2), \quad (5, 3),$$
$$(1, 2), \quad (1, 3), \quad (2, 5), \quad (3, 5).$$

## Solutions to Testing Questions    27

### Testing Questions    (27-A)

1.  $\dfrac{c}{a+b} < \dfrac{a}{b+c} < \dfrac{b}{a+c} \Rightarrow \dfrac{c}{a+b} + 1 < \dfrac{a}{b+c} + 1 < \dfrac{b}{a+c} + 1,$

    i.e. $\dfrac{1}{a+b} < \dfrac{1}{b+c} < \dfrac{1}{a+c}$, so

    $$a+b > b+c > a+c, \ \therefore c < a < b.$$

2.  $a < b < c < 0$ gives $-a > -b > -c > 0$, so that

    $$0 < -(b+c) < -(c+a) < -(a+b),$$
    $$\therefore \frac{-a}{-(b+c)} > \frac{-b}{-(c+a)} > \frac{-c}{-(a+b)},$$

    which implies that $\dfrac{a}{b+c} > \dfrac{b}{c+a} > \dfrac{c}{a+b}.$

3.  The given inequality yields

    $$(a-1)x < b-4.$$

    (i)     When $a > 1$, then $x < \dfrac{b-4}{a-1}.$

    (ii)    When $a < 1$, then $x > \dfrac{b-4}{a-1}.$

    (iii)   When $a = 1$ and $b > 4$, then $0 \cdot x < b - 4$, solution set $= \mathbb{R}_1.$

    (iv)    When $a = 1$ and $b \leq 4$, then no solution.

4.  $m > n$ leads to $\dfrac{4-x}{3} > \dfrac{x+3}{4}$, i.e. $16 - 4x > 3x + 9$, so $x < 1.$

    $n > p$ leads to $\dfrac{x+3}{4} > \dfrac{2-3x}{5}$, i.e. $5x + 15 > 8 - 12x$, so $x > -\dfrac{7}{17}.$

    Thus, the range of $x$ is $-\dfrac{7}{17} < x < 1.$

5.
$$\begin{cases} x - 1 > -3 \\[2mm] \dfrac{1}{2}x - 1 < \dfrac{1}{3}x \\[2mm] 3 < 2(x-1) < 10 \\[2mm] \dfrac{1}{3}(3-2x) > -2 \end{cases} \Rightarrow \begin{cases} x > -2 \\[2mm] \dfrac{1}{6}x < 1 \\[2mm] \dfrac{3}{2} < (x-1) < 5 \\[2mm] 3 - 2x > -6 \end{cases} \Rightarrow \begin{cases} x > -2 \\[2mm] x < 6 \\[2mm] \dfrac{5}{2} < x < 6 \\[2mm] \dfrac{9}{2} > x \end{cases}$$

Thus, the solution set is $\{\dfrac{5}{2} < x < \dfrac{9}{2}\}$.

6.  The system $\dfrac{x}{y} = \dfrac{a}{b}, x + y = c$ yields $x = \dfrac{ac}{a+b}, y = \dfrac{bc}{a+b}$. Since $c > 0$,
so $x < y$, the answer is (D).

7.  The given conditions implies that $2a - b < 0$ and $\dfrac{a - 2b}{2a - b} = \dfrac{5}{2}$, so

$$2(a - 2b) = 5(2a - b), \quad \text{i.e. } b = 8a.$$
$$\therefore 2a - b = -6a < 0, \quad \text{i.e. } a > 0.$$

Now $ax + b < 0$ yields $x < -\dfrac{b}{a} = -8$, so the solution set of the inequality
$ax + b < 0$ is

$$x < -8.$$

8.  The given conditions implies that $a > 0$, so $1 + \dfrac{b}{a} + \dfrac{c}{a} = 0$ and $1 > \dfrac{b}{a} > \dfrac{c}{a}$.
$\dfrac{b}{a} = -1 - \dfrac{c}{a}$ implies $1 > -1 - \dfrac{c}{a} > \dfrac{c}{a}$, so

$$-2 < \dfrac{c}{a} < -\dfrac{1}{2}.$$

9.  From $(2a - b)x + a - 5b > 0$,

$$(2a - b)x > 5b - a,$$
$$x < \dfrac{5b - a}{2a - b}, \quad \text{where } 2a - b < 0.$$

Thus, $2a < b$ and $\dfrac{5b - a}{2a - b} = \dfrac{10}{7}$, so

$$7(5b - a) = 10(2a - b),$$
$$35b - 7a = 20a - 10b,$$

$$a = \dfrac{45b}{27} = \dfrac{5}{3}b.$$

$2a - b < 0$, implies $\dfrac{10}{3}b - b < 0$, so $a, b < 0$, hence the solution set $ax > b$ is $x < \dfrac{b}{a} = \dfrac{3}{5}$, i.e. $x < \dfrac{3}{5}$.

10. The solution set of the system is the set of the integers $x$ satisfying $\dfrac{a}{9} \le x < \dfrac{b}{8}$. Since each of $1, 2, 3$ satisfies the two inequalities,

$$0 < \frac{a}{9} \le 1 \Rightarrow 0 < a \le 9,$$
$$3 < \frac{b}{8} \le 4 \Rightarrow 24 < b \le 32,$$

so $a$ has 9 choices and $b$ has 8 choices, i.e. there are $9 \times 8 = 72$ required ordered pairs $(a, b)$.

## Testing Questions (27-B)

1. It is needed to represent $a - 2b$ in terms of $a - b$ and $a + b$. Since

$$a = \frac{1}{2}[(a - b) + (a + b)], \quad -2b = (a - b) - (a + b),$$

so $a - 2b = -\frac{1}{2}(a + b) + \frac{3}{2}(a - b)$. $a - 2b$ will take its maximum value if $a + b = 1, a - b = 1$, i.e. $a = 1, b = 0$. Thus,

$$8a + 2002b = 8a = 8.$$

2. Label the equations as

$$3x + 2y - z = 4, \quad (30.20)$$
$$2x - y + 2z = 6, \quad (30.21)$$
$$x + y + z < 7. \quad (30.22)$$

By (30.20) $+2\times$ (30.21),

$$7x + 3z = 16,$$
$$\therefore x = 1, z = 3.$$

From (30.21), $y = 2x + 2z - 6 = 2$. Since $x = 1, y = 2, z = 3$ satisfy the inequality (30.22), the solution is $x = 1, y = 2, z = 3$.

3. By adding up the given two inequalities, (C) is obtained.

   When $x = y = 2, z = 1$, then (A) is not true. When $x = y = 1, z = 0.7$, then (B) is not true. When $x = y = -1, z = -1.5$, then (D) is not rue. Thus, only (C) is always true.

4. The inequalities $0 \leq ax + 5 \leq 4$ yields $-5 \leq ax \leq -1$. Since the integer solutions for $x$ are positive, so $a < 0$. Therefore

$$-\frac{1}{a} \leq x \leq -\frac{5}{a}.$$

   Since $0 < -\frac{1}{a} \leq 1$ and $4 \leq -\frac{5}{a} < 5$, so

$$a \leq -1 \quad \text{and} \quad a \geq -\frac{5}{4}.$$

   Thus, the range of $a$ is $-\frac{5}{4} \leq a \leq -1$.

5. The given conditions $\frac{8}{9} < \frac{a}{b} < \frac{9}{10}$ implies

$$8b < 9a \quad \text{and} \quad 10a < 9b,$$
$$8b + 1 \leq 9a, \quad \text{and} \quad 10a + 1 \leq 9b,$$
$$\therefore 8b + 1 \leq 9 \cdot \frac{9b - 1}{10},$$
$$80b + 10 \leq 81b - 9,$$
$$b \geq 19.$$

   When let $b = 19, a = \frac{9 \cdot 19 - 1}{10} = 17$, then $\frac{a}{b} = \frac{17}{19}$.

$$\because \frac{8}{9} < \frac{17}{19} < \frac{9}{10},$$

   and $b$ is the minimum possible value, $\frac{17}{19}$ is the fraction to be found.

# Solutions to Testing Questions    28

## Testing Questions    (28-A)

1. Let $S$ be the sign of $(2 + x)(x - 5)(x + 1)$, then we have

| Range of $x$ | $x < -2$ | $-2 < x < -1$ | $-1 < x < 5$ | $5 < x$ |
|---|---|---|---|---|
| $S$ | $-$ | $+$ | $-$ | $+$ |

Therefore the solution set is $\{-2 < x < -1\} \cup \{5 < x\}$.

2. The given inequality implies that $x \neq 0$, so $x^2(x^2 - 4) < 0 \Leftrightarrow x^2 - 4 < 0$, hence the solution set is $\{-2 < x < 0\} \cup \{0 < x < 2\}$.

3. $x^3 + x^2 - 6x \leq 0 \Leftrightarrow x(x^2 + x - 6) \leq 0 \Leftrightarrow x(x+3)(x-2) \leq 0$. Let $S$ be the sign of $x(x+3)(x-2)$, then

| Range of $x$ | $x < -3$ | $-3 < x < 0$ | $0 < x < 2$ | $2 < x$ |
|---|---|---|---|---|
| $S$ | $-$ | $+$ | $-$ | $+$ |

Thus, by adding the points $-3, 0$ and $2$, the solution set is $\{x \leq -3\} \cup \{0 \leq x \leq 2\}$.

4. $x - 1 > (x-1)(x+2) \Leftrightarrow (x-1)(x+1) < 0$ i.e. $x^2 - 1 < 0$, therefore the solution set is $|x| < 1$ i.e. $\{-1 < x < 1\}$.

5. $(x^3 - 1)(x^3 + 1) > 0 \Leftrightarrow x^6 - 1 > 0 \Leftrightarrow x^6 > 1$, so the solution set is $|x| > 1$ or equivalently, $\{x < -1\} \cup \{x > 1\}$.

6. $\dfrac{2x - 4}{x + 3} > \dfrac{x + 2}{2x + 6} \Leftrightarrow \dfrac{2(2x-4) - (x+2)}{2(x+3)} > 0 \Leftrightarrow \dfrac{3x - 10}{2(x+3)} > 0,$

   When $3x - 10 > 0$ and $2(x+3) > 0$, then $x > \dfrac{10}{3}$.

   When $3x - 10 < 0$ and $x + 3 < 0$, then $x < -3$, therefore the solution set is $\{x < -3\} \cup \{x > \dfrac{10}{3}\}$.

7. $\dfrac{x}{x + 2} \geq \dfrac{1}{x}$ implies that $x \neq 0, -2$ and

$$\frac{x}{x + 2} \geq \frac{1}{x} \Leftrightarrow \frac{x^2 - (x + 2)}{x(x + 2)} \geq 0 \Leftrightarrow \frac{(x - 2)(x + 1)}{x(x + 2)} \geq 0.$$

Let $S$ be the sign of $\dfrac{(x - 2)(x + 1)}{x(x + 2)}$, then

| Range of $x$ | $(-\infty, -2)$ | $(-2, -1)$ | $(-1, 0)$ | $(0, 2)$ | $(2, +\infty)$ |
|---|---|---|---|---|---|
| $S$ | $+$ | $-$ | $+$ | $-$ | $+$ |

so the solution set is $(-\infty, -2) \cup [-1, 0) \cup [2, +\infty)$.

8.  The given inequality $\dfrac{x-1}{x^2} \le 0$ implies that $x \ne 0$, and the inequality is
    equivalent to $x - 1 \le 0$, therefore the solution set is

    $$(-\infty, 0) \cup (0, 1].$$

9.  The inequality $\dfrac{x(2x-1)^2}{(x+1)^3(x-2)} > 0$ implies that $x \ne -1, 0, \frac{1}{2}, 2$, and the
    inequality is equivalent to

    $$\frac{x}{(x+1)(x-2)} > 0.$$

    Let $S$ be the sign of the value of $\dfrac{x}{(x+1)(x-2)}$, then

    | Range of $x$ | $x < -1$ | $-1 < x < 0$ | $0 < x < 2$ | $2 < x$ |
    |---|---|---|---|---|
    | $S$ | $-$ | $+$ | $-$ | $+$ |

    therefore the solution set is

    $$(-1, 0) \cup (2, +\infty).$$

10. From the given inequality we have $x \ne -1$ and $0$ is in the solution set. Then
    let $x \ne 0$ first, so that

    $$\frac{2x^2}{x+1} \ge x \Leftrightarrow \frac{2x^2 - x(x+1)}{x(x+1)} \ge 0 \Leftrightarrow \frac{x(x-1)}{x(x+1)} \ge 0 \Leftrightarrow \frac{x-1}{x+1} \ge 0.$$

    If $x - 1 \ge 0$ and $x + 1 > 0$ then $x \ge 1$; if $x - 1 \le 0$ and $x + 1 < 0$ then
    $x < -1$. Thus the solution set is $\{x < -1\} \cup \{0\} \cup \{x \ge 1\}$.

## Testing Questions    (28-B)

1.  Let $y = x^2$, then $y \ne 1$ or $3$, annd the given inequality becomes

    $$\frac{y+3}{y+1} + \frac{y-5}{y-3} \ge \frac{y+5}{y+3} + \frac{y-3}{y-1}$$
    $$\left(1 + \frac{2}{y+1}\right) + \left(1 - \frac{2}{y-3}\right) \ge \left(1 + \frac{2}{y+3}\right) + \left(1 - \frac{2}{y-1}\right),$$
    $$\frac{1}{y+1} - \frac{1}{y+3} \ge \frac{1}{y-3} - \frac{1}{y-1},$$

therefore

$$\frac{1}{(y+1)(y+3)} \geq \frac{1}{(y-3)(y-1)}. \qquad (*)$$

When $0 < y < 1$ or $y > 3$, then $(y+1)(y+3) > 0$ and $(y-3)(y-1) > 0$, so $(*)$ becomes $(y+1)(y+3) \leq (y-3)(y-1)$, i.e. $8y \leq 0$, so no solution. When $1 < y < 3$, then $(y-3)(y-1) < 0$ but $(y+1)(y+3) > 0$, so $(*)$ holds. Returning to $x$, the solution set in $\{x > 0\}$ is $\{1 < x < \sqrt{3}\}$.

2. Both sides of the given inequality are multiplied by 4, it follows that

$$\left(1+\frac{5}{4x+3}\right)-\left(1-\frac{1}{4x+1}\right) > \left(1+\frac{1}{4x-1}\right)-\left(1-\frac{5}{4x-3}\right),$$

$$\frac{5}{4x+3}-\frac{5}{4x-3} > \frac{1}{4x-1}-\frac{1}{4x+1},$$

$$\frac{15}{16x^2-9} < \frac{-1}{16x^2-1}.$$

When $16x^2 - 9 > 0$ then $16x^2 - 1 > 0$, so right hand side of the last inequality is negative, no solution.

When $16x^2 - 9 < 0$ and $16x^2 - 1 < 0$, then the last inequality holds, the solution set is $\{|x| < \frac{1}{4}\}$.

When $16x^2 - 9 < 0$ and $16x^2 - 1 > 0$, then $15(16x^2 - 1) > -(16x^2 - 9)$, so $x^2 > \frac{3}{32}$, the solution set is $\{|x| > \frac{\sqrt{3}}{4\sqrt{2}}\}$.

Thus, the solution set is $\{|x| < \frac{1}{4}\} \cup \{|x| > \frac{\sqrt{3}}{4\sqrt{2}}\}$.

3. Let $f(x) = a^3 + b^3 - x^3 - (a+b-x)^3 - m$, then $f(x) \leq 0$ for any real $x$, and

$$f(x) = -3ab(a+b) + 3(a+b)^2 x - 3(a+b)x^2 - m$$

$$= -3(a+b)[x^2 - (a+b)x + \frac{1}{4}(a+b)^2] + \frac{3}{4}(a+b)[(a+b)^2 - 4ab] - m$$

$$= -3(a+b)\left(x - \frac{a+b}{2}\right)^2 + \frac{3}{4}(a+b)(a-b)^2 - m.$$

Since the maximum value of $-3(a+b)\left(x - \frac{a+b}{2}\right)^2$ is 0, so

$$\frac{3}{4}(a+b)(a-b)^2 - m \leq 0,$$

$$\therefore m \geq \frac{3}{4}(a+b)(a-b)^2.$$

Thus, the minimum value of $m$ is $\frac{3}{4}(a+b)(a-b)^2$.

4.  For $a < -2$, The condition $f(-2) \geq a$ yields $4a + 10 \geq a$ i.e. $-\frac{10}{3} \leq a <$
    $-2$. Since the symmetric axis of the curve $y = f(x)$ is $x = a$, so if $a < -2$,
    then $f(x) \geq f(-2)$ for $x \geq -2$. Thus, the requirement is satisfied.

    For $-2 \leq a \leq 2$, $f(x) = (x-a)^2 + 6 - a^2 \geq 6 - a^2 \geq 2 \geq a$ for any real
    $x$.

    For $a > 2$, $f(2) \geq a$ yields $4 - 4a + 6 \geq a$ i.e. $a \leq 2$, a contradiction, so
    the range of $a$ is $-\frac{10}{3} \leq a \leq 2$.

5.  Let $y = x^2 - 3x + 2 = (x - \frac{3}{2})^2 - \frac{1}{4}$. Then $y_{min} = -\frac{1}{4}$ and $y_{max} = 2$ as
    $x = \frac{3}{2}$ and $x = 0$ respectively. Therefore

    $$\frac{1}{8}(2a - a^2) \leq -\frac{1}{4}, \quad 2 \leq 3 - a^2,$$
    $$a^2 - 2a - 2 \geq 0, \quad a^2 \leq 1,$$
    $$|a - 1| \geq \sqrt{3}, \quad |a| \leq 1,$$

    thus, the range of $a$ is $\left\{ \{a \leq 1 - \sqrt{3}\} \cup \{a \geq 1 + \sqrt{3}\} \right\} \cap \{-1 \leq a \leq 1\} =$
    $\{-1 \leq a \leq 1 - \sqrt{3}\}$.

# Solutions to Testing Questions    29

## Testing Questions    (29-A)

1.  $|x^2 + x + 1| \leq 1 \Leftrightarrow -1 \leq x^2 + x + 1 \leq 1 \Leftrightarrow x^2 + x + 2 \geq 0$ and $x^2 + x \leq 0$.

    Since $x^2 + x + 2 = (x + \frac{1}{2})^2 + \frac{7}{4} > 0$ for each real $x$, the solution set is
    $\mathbb{R}$.

    $x^2 + x \leq 0 \Rightarrow x(x + 1) \leq 0$, its solution set is $-1 \leq x \leq 0$.

    Thus, the solution set to the original inequality is $\mathbb{R} \cap \{-1 \leq x \leq 0\} =$
    $\{-1 \leq x \leq 0\}$.

2.  The given inequality is equivalent to $(3 - 2x)^2 \le (x + 4)^2$, therefore

$$9 - 12x + 4x^2 \le x^2 + 8x + 16,$$
$$3x^2 - 20x - 7 \le 0,$$
$$(3x + 1)(x - 7) \le 0,$$
$$\therefore -\frac{1}{3} \le x \le 7.$$

Thus the solution set is $\{-\frac{1}{3} \le x \le 7\}$.

3.  It is clear that $x - 1 \ne 0$ i.e. $x \ne 1$, so $|x - 1| > 0$ and

$$\left|\frac{x + 1}{x - 1}\right| \ge 1 \Leftrightarrow |x + 1| \ge |x - 1| \Leftrightarrow (x + 1)^2 \ge (x - 1)^2,$$

i.e.,

$$x^2 + 2x + 1 \ge x^2 - 2x + 1,$$
$$4x \ge 0, \therefore x \ge 0.$$
$$\because x \ne 1,$$

the solution set is $(0, 1) \cup (1, +\infty)$.

4.  $|x + 3| > 2x + 3 \Leftrightarrow x + 3 < -(2x + 3)$ or $x + 3 > 2x + 3 \Leftrightarrow x < -2$ or $x < 0$. Thus, the solution set is $(-\infty, -2) \cup (-\infty, 0) = (-\infty, 0)$.

5.  $|x^2 - 4x - 5| > x^2 - 4x - 4 \Leftrightarrow x^2 - 4x - 5 < -(x^2 - 4x - 4)$ or $x^2 - 4x - 5 > x^2 - 4x - 4$. The second inequality has no real solution, and $x^2 - 4x - 5 < -(x^2 - 4x - 4)$ yields

$$2x^2 - 8x - 9 < 0,$$
$$\therefore 2 - \tfrac{1}{2}\sqrt{34} < x < 2 + \tfrac{1}{2}\sqrt{34},$$

i.e. the solution set is $\{2 - \tfrac{1}{2}\sqrt{34} \le x \le 2 + \tfrac{1}{2}\sqrt{34}\}$.

6.  Since $x \ne 0$ and any $x < 0$ satisfies the given inequality, the set $(-\infty, 0)$ is a part of the solution set.

For $x > 0$, then $|x + 1| > \dfrac{2}{x} \Leftrightarrow x^2 + x > 2 \Leftrightarrow (x + 2)(x - 1) > 0$, therefore $1 < x$. Thus, the solution set is $\{x < 0\} \cup \{1 < x\}$.

7.  For $x \le -1$, the given inequality becomes $-(x + 1) + (2 - x) \le 3x$, therefore $x \ge \frac{1}{5}$, but it is not acceptable, so no solution.

For $-1 \le x \le 2$, the given inequality becomes $(x + 1) + (2 - x) \le 3x$, therefore the set $1 \le x \le 2$ is in the solution set.

For $2 < x$, the given inequality becomes $(x + 1) + (x - 2) \le 3x$, therefore $2 < x$ is in the solution set.

Thus, the solution set is $\{1 \le x\}$.

8.  (i) Let $y = |x|$, then $y^2 + y - 6 > 0$, so $(y + 3)(y - 2) > 0$, the solution set for $y$ is $2 < y$ (since $y \ge 0$). Returning to $x$, the solution set for $x$ is $\{|x| > 2\}$, or equivalently, $\{x < -2\} \cup \{x > 2\}$.

    (ii) Let $y = |x|$, then $y \ge 0$ and $y \ne 2$. Hence the given inequality is equivalent to

    $$\frac{y - 1}{y - 2} < 0,$$
    $$(y - 1)(y - 2) < 0,$$
    $$1 < y < 2, \quad \text{i.e. } 1 < |x| < 2.$$

    Thus the solution set is $\{-2 < x < -1\} \cup \{1 < x < 2\}$.

9.  For $x > 0$, the given inequality becomes $|x^2 - 1| > 1$, i.e. $x^2 - 1 > 1$ or $x^2 - 1 < -1$ (no solution), so the solution set is $x > \sqrt{2}$.

    For $x < 0$, the given inequality becomes $|x^2 - 1| > -1$, so the solution set is any negative number.

    Thus the solution set of the question is $\{x < 0\} \cup \{x > \sqrt{2}\}$.

10. By taking squares to both sides, the absolute value signs in the outer layer can be removed.

    $$||a| + (a - b)| > |a + |a - b|| \Leftrightarrow (|a| + (a - b))^2 > (a + |a - b|)^2,$$
    $$a^2 + (a - b)^2 + 2|a|(a - b) > a^2 + (a - b)^2 + 2a|a - b|,$$
    $$|a|(a - b) > a|a - b|,$$

    therefore $a, a - b$ are both not zero, so $\dfrac{a}{|a|} < \dfrac{a - b}{|a - b|}$. Since both sides have absolute value 1, so $a < 0$ and $a - b > 0$, thus, $a < 0, b < 0$, the answer is (D).

## Testing Questions (29-B)

1.  (i) By symmetry we may suppose that $a \ge b \ge c$. Then $a > 0$ and $b + c = 2 - a, bc = \dfrac{4}{a}$. so $b, c$ are the real roots of the quadratic equation

    $$x^2 - (2 - a)x + \frac{4}{a} = 0.$$

Its discriminant $\Delta \geq 0$ implies $(2 - a)^2 - \dfrac{16}{a} \geq 0$, so

$$a^3 - 4a^2 + 4a - 16 \geq 0,$$
$$(a - 4)(a^2 + 4) \geq 0,$$
$$\therefore a - 4 \geq 0 \text{ i.e. } a \geq 4.$$

In fact, when $a = 4, b = c = -1$, the conditions are satisfied. Thus, the minimum value of the maximal value of $a, b, c$ is 4.

(ii) The given conditions implies that $a, b, c$ may be all positive or one positive and two negative. Let $a > 0, b < 0, c < 0$. then

$$|a| + |b| + |c| = a - b - c = 2a - 2 \geq 8 - 2 = 6,$$

so the minimum value of $|a| + |b| + |c|$ is 6.

2.  The condition $b = \dfrac{a + c}{2}$ implies $2b = a + c$ or $a - b = b - c$, so $|a - b| = |b - c|$, Since $|a| < |c|$, so

$$S_1 = \left|\frac{a - b}{c}\right| = \left|\frac{b - c}{c}\right| < \left|\frac{b - c}{a}\right| = S_2.$$

$2b = a + c$ implies $2(b - c) = a - c$, so $2|b - c| = |a - c|$. Since $2|a| > |b|$, so

$$S_2 = \left|\frac{b - c}{a}\right| = \left|\frac{2(b - c)}{2a}\right| < \left|\frac{a - c}{b}\right| = S_3.$$

Thus, $S_1 < S_2 < S_3$, the answer is (A).

3.  For $y \geq 0$, the given inequality becomes $\dfrac{3y + 10}{8y + 5} > 1$, and which is equivalent to

$$3y + 10 > 8y + 5 \Leftrightarrow y < 1.$$

Thus, the solution set is $0 \leq y < 1$ for $y \geq 0$.

For $y < 0$, then given inequality becomes $\dfrac{|y + 10|}{|4y + 5|} > 1$, so $y \neq -\dfrac{5}{4}$ or $-10$.

For $y < -10$, $\dfrac{|y + 10|}{|4y + 5|} > 1 \Leftrightarrow -(y + 10) > -(4y + 5) \Leftrightarrow y > \dfrac{5}{3}$, no solution.

For $-10 < y < -\dfrac{5}{4}$, $\dfrac{|y + 10|}{|4y + 5|} > 1 \Leftrightarrow 10 + y > -(4y + 5) \Leftrightarrow y > -3$,

so the solution set is $-3 < y < -\dfrac{5}{4}$.

For $-\dfrac{5}{4} < y < 0$, $\dfrac{|y+10|}{|4y+5|} > 1 \Leftrightarrow y + 10 > 4y + 5 \Leftrightarrow y < \dfrac{5}{3}$, so the solution set is $-\dfrac{5}{4} < y < 0$.

By adding the parts of solution set, the solution set for the original inequality is

$$\{-3 < y < -\tfrac{5}{4}\} \cup \{-\tfrac{5}{4} < y < 1\}.$$

4.  The condition (i) implies that the curve $y = ax^2 + bx + x$ is open upwards. The conditions (i) and (iii) imply that

$$a + b = 2. \tag{30.23}$$

The condition (ii) implies that

$$|a + b + c| \le 1, \tag{30.24}$$

$$|c| \le 1. \tag{30.25}$$

(30.23) and (30.24) implies that $|2 + c| \le 1$ i.e.

$$-3 \le c \le -1, \tag{30.26}$$

(30.25) and (30.26) implies that $c = -1$. Thus, the curve $y = ax^2 + bx + c$ reaches its minimum value $-1$ at $x = 0$, so $-\dfrac{b}{2a} = 0$ i.e. $b = 0$, which implies that $a = 2$. Thus, $a = 2, b = 0, c = -1$.

5.  Let $a = b = -\sqrt[3]{2}, c = \dfrac{1}{2} \cdot \sqrt[3]{2}$, then $a, b, c$ satisfy all the conditions in question, so $k \le 4$.

Below we show that the inequality $|a + b| \ge 4|c|$ holds for any $(a, b, c)$ which satisfies all conditions in question.

The given conditions implies $a, b, c$ are all non-zero and $c > 0$, and

$$ab = \dfrac{1}{c} > 0, \quad 0 = ab + bc + ca = \dfrac{1}{c} + (a+b)c \Leftrightarrow a + b = -\dfrac{1}{c^2} < 0,$$

so $a \le b < 0$. By inverse Viete Theorem, $a, b$ are the roots of the quadratic equation

$$x^2 + \dfrac{1}{c^2}x + \dfrac{1}{c} = 0,$$

so $\Delta = \dfrac{1}{c^4} - \dfrac{4}{c} \ge 0$, i.e. $c^3 \le \dfrac{1}{4}$. Thus,

$$|a + b| = -(a + b) = \dfrac{1}{c^2} \ge 4c = 4|c|.$$

# Solutions to Test Questions   30

## Testing Questions   (30-A)

1. Based on Example 1, it is obtained that $AB + AC > RB + RC$,

   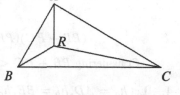

   $BA + BC > RC + RA$, and
   $BC + CA > RA + RB$.

   Adding them up, then

   $2(AB+BC+CA) > 2(RA+RB+RC)$,

   so

   $$RA + RB + RC < AB + BC + CA.$$

   From triangle inequality, $RA + RB > AB$, $RB + RC > BC$, $RC + RA > CA$. Adding them up, the inequality

   $$\frac{1}{2}(AB + BC + CA) < RA + RB + RC$$

   is obtained at once.

2. Since $\angle C > \angle B$, we have $AB > AC$. On $AB$ take $C'$ such that $AC' = AC$,

   and make $C'D \perp BE$ at $D$ and
   $C'F' \parallel BE$, intersecting $AC$ at $F'$, as
   shown in the diagram, then $DC'F'E$
   is a rectangle and

   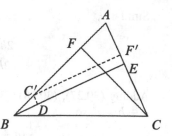

   $$CF = C'F' = DE.$$

   Therefore $AB + CF = AC' + BC' + C'F' = AC + BC' + DE$, so that

   $$(AB + CF) - (AC + BE) = BC' - BD > 0$$

   since $BC'$ is the hypotenuse in the Rt$\triangle BC'D$. Thus, $AB + CF > AC + BE$.

3. By Pythagoras' Theorem, $PB^2 - PC^2 = (PD^2 + BD^2) - (PD^2 + DC^2)$

$$= BD^2 - DC^2 \quad \text{and}$$

$$AB^2 - AC^2$$
$$= (AD^2 + BD^2) - (AD^2 + DC^2)$$
$$= BD^2 - DC^2,$$

so
$$PB^2 - PC^2 = AB^2 - AC^2. \text{ Thus,}$$

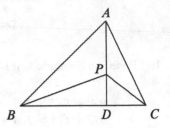

$$(PB - PC)(PB + PC) = (AB - AC)(AB + AC).$$

Considering $PB + PC < AB + AC$, we obtain $PB - PC > AB - AC$.

4.  Let $h_a = AD, h_b = BE, h_c = CF$, Then

$$h_a < b, h_a < c, h_b < a, h_b < c, h_c < b, h_c < a.$$

Adding them up, we obtain $2(h_a + h_b + h_c) < 2(a + b + c)$, i.e.

$$h_a + h_b + h_c < a + b + c.$$

On the other hand, From $AD + BD > AB, AD + DC > AC$ we have

$$2h_a + a > b + c. \tag{30.27}$$

Similarly,

$$2h_b + b > c + a, \tag{30.28}$$
$$2h_c + c > a + b. \tag{30.29}$$

Adding (30.27), (30.28), (30.29) up, it follows that $2(h_a + h_b + h_c) > a + b + c$, so
$$\frac{1}{2}(a + b + c) < h_a + h_b + h_c.$$

5.  The assumptions implies that $a + b > c, \ b + c > a, \ c + a > b$. Since

$$\frac{1}{c+a} > \frac{1}{a+b+a+b} = \frac{1}{2(a+b)} \quad \text{and}$$
$$\frac{1}{b+c} > \frac{1}{a+b+a+b} = \frac{1}{2(a+b)},$$

it follows that $\dfrac{1}{b+c} + \dfrac{1}{c+a} > \dfrac{1}{a+b}$. Similarly,

$$\frac{1}{a+b} + \frac{1}{c+a} > \frac{1}{b+c} \quad \text{and} \quad \frac{1}{a+b} + \frac{1}{b+c} > \frac{1}{a+c}.$$

The three inequalities prove the conclusion.

6. Since

$$2(u^2 + v^2) \ge (u+v)^2$$

for any real numbers $u$ and $v$, and

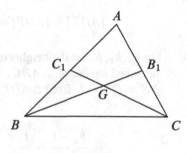

$$BB_1 = \frac{3}{2}BG, CC_1 = \frac{3}{2}CG,$$

it follows that

$$BB_1^2 + CC_1^2 = \frac{9}{4}(BG^2 + CG^2)$$
$$\ge \frac{9}{8}(BG + CG)^2 > \frac{9}{8}BC^2.$$

7. Let $MB = a, BK = b, KC = c, AC = d, AM = e$. Then

$$[MBK] > [MCK] \Longrightarrow b > c,$$
$$[MBK] > [MAK] \Longrightarrow a > e.$$

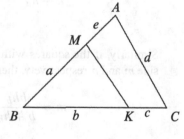

Suppose that $\dfrac{a+b}{c+d+e} < \dfrac{1}{3}$,

then $3a + 3b < c + d + e < b + d + a$,

so $2a + 2b < d$.

Since $2a + 2b > (a + e) + (b + c) = AB + BC$, so $AB + BC < d = AC$,

a contradiction. Thus, the conclusion is proven.

8. The $n$ lines can form $4\binom{n}{2} = 2n(n-1)$ angles. If by translation we move all the lines such that they are all pass a fixed point $O$ in the plane, they form $2n$ angles, and each is one of the $2n(n-1)$ angles.

If each of the $2n$ angles is greater than $\dfrac{180°}{n}$, then their sum is greater than $2n \cdot \dfrac{180°}{n} = 360°$, a contradiction. Thus, the conclusion is proven.

9. Extending $ED$ to $E_1$ such that $DE_1 = DE$. Connect $E_1B, E_1F$. Then

$\triangle BDE_1 \cong \triangle ADE$ (S.A.S.).

For the quadrilateral $E_1BFD$, we have

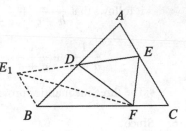

$$[FDE_1] = [DEF],$$

so that

$$[ADE] + [BDF] = [E_1BFD] > [FDE_1] = [DEF].$$

10. Let $h_a, h_b, h_c$ be the heights on $BC, CA, AB$ respectively. If $PQRS$ is an inscribed square of $\triangle ABC$, such that $RS$ is on $BC$, Let $PQ = PS = RS = QR = l$, from $\triangle AQP \sim \triangle ABC$, we have

$$\frac{h_a - l}{h_a} = \frac{l}{a}$$

we have

$$l = \frac{ah_a}{a + h_a}.$$

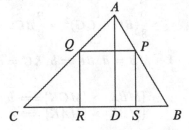

Similarly, if the squares with one side on $AC$ and on $AB$ have length of side $m$ and $n$ respectively, then

$$m = \frac{bh_b}{b + h_b} \qquad \text{and} \qquad n = \frac{ch_c}{c + h_c}.$$

Since $\dfrac{b}{a} = \dfrac{h_a}{h_b}$, so $\dfrac{a - h_b}{b - h_a} = \dfrac{a}{b} > 1$, i.e. $a - h_b > b - h_a$, therefore $a + h_a > b + h_b$. Thus,

$$\frac{ah_a}{a + h_a} < \frac{bh_b}{b + h_b}, \qquad \therefore l < m.$$

Similarly, $m < n$. Thus, the square with one side on the shortest side $AB$ has maximum area.

## Test Questions   (30-B)

1. (KIEV/1969) Suppose that there is such a triangle. Let its area be $S$. Then we can assume that

$$a = \frac{2S}{1}, \qquad b = \frac{2S}{\sqrt{5}}, \qquad c = \frac{2S}{1 + \sqrt{5}}.$$

Since $a > b > c$, we check if the triangle inequality holds for the triangle, and it suffices to check $b + c > a$.

$$b + c = 2S \left( \frac{1}{\sqrt{5}} + \frac{1}{1 + \sqrt{5}} \right) < 2S \left( \frac{1}{2} + \frac{1}{1 + 2} \right) = 2S \cdot \frac{5}{6} < a,$$

a contradiction. Thus, such a triangle does not exist.

2. First of all we have

$$A_1 C - C B_1 < A_1 B_1, \quad B_1 A - A C_1 < B_1 C_1,$$
$$C_1 B - B A_1 < C_1 A_1,$$

i.e. $\frac{3}{4} a - \frac{1}{4} b < c_1, \frac{3}{4} b - \frac{1}{4} c < a_1,$
$\frac{3}{4} c - \frac{1}{4} a < b_1,$

where $a, b, c$ are the lengths of $BC, CA, AB$ respectively, and $a_1, b_1, c_1$ are the lengths of $B_1 C_1, C_1 A_1, A_1 B_1$ respectively. By adding them up we obtain

$$\frac{1}{2}(a + b + c) < a_1 + b_1 + c_1, \qquad \text{i.e.} \quad \frac{1}{2} P < p.$$

On the sides of $\triangle ABC$ we take segments $A_1 A_2, B_1 B_2, C_1 C_2$ such that $A_1 A_2 = \frac{1}{2} a, B_1 B_2 = \frac{1}{2} b, C_1 C_2 = \frac{1}{2} c$. It is easy to see that $B_2 C_1 = \frac{1}{4} a, A_1 C_2 = \frac{1}{4} b, A_2 B_1 = \frac{1}{4} c$, so that

$$\frac{1}{2} b + \frac{1}{4} a > a_1, \quad \frac{1}{2} c + \frac{1}{4} b > b_1, \quad \frac{1}{2} a + \frac{1}{4} c > c_1.$$

Adding them up, we obtain

$$a_1 + b_1 + c_1 < \frac{3}{4}(a + b + c),$$

i.e. $p < \frac{3}{4} P$.

3.

$$\begin{aligned}
I^2 - 3S &= (a+b+c)^2 - 3(ab+bc+ca) \\
&= a^2 + b^2 + c^2 - ab - bc - ca \\
&= \tfrac{1}{2}[(a-b)^2 + (b-c)^2 + (c-a)^2] \geq 0,
\end{aligned}$$

therefore $3S \leq I^2$. On the other hand,

$$\begin{aligned}
I^2 - 4S &= (a+b+c)^2 - 4(ab+bc+ca) \\
&= a^2 + b^2 + c^2 - 2ab - 2bc - 2ca \\
&= (a-b)^2 + c^2 - 2c(a+b) \\
&< c^2 + c^2 - 2c(a+b) = 2c[c-(a+b)] < 0,
\end{aligned}$$

therefore $I^2 < 4S$.

4.  From $1 = (a+b+c)^2 = a^2 + b^2 + c^2 + 2(ab+bc+ca)$, we have $4(ab+bc+ca) = 2 - 2(a^2+b^2+c^2)$. From Heron's formula, the area $S$ of the triangle is given by

$$S = \sqrt{\frac{1}{2}\left(\frac{1}{2}-a\right)\left(\frac{1}{2}-b\right)\left(\frac{1}{2}-c\right)},$$

so that

$$\begin{aligned}
16S^2 &= (1-2a)(1-2b)(1-2c) \\
&= 1 - 2(a+b+c) + 4(ab+bc+ca) - 8abc \\
&= -1 + 4(ab+bc+ca) - 8abc \\
&= 1 - 2(a^2+b^2+c^2) - 8abc,
\end{aligned}$$

therefore $1 - 2(a^2+b^2+c^2) - 8abc > 0$, i.e. $a^2+b^2+c^2+4abc < \dfrac{1}{2}$.

5.  (i)  When $X, Y, Z$ are midpoints of corresponding sides, then

$$[XYZ] = \frac{1}{4}[ABC].$$

If $X$ is not the midpoint of $BC$. Since $BX < \frac{1}{2}BC$, when $Y, Z$ are fixed and moving $X$ to the midpoint of $BC$, the the height of $\triangle XYZ$ on the side $YZ$ is reduced, so $[XYZ]$ is reduced. In similar way, we can also move $Y, Z$ to the midpoint if needed. Therefore we have

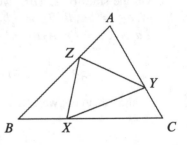

$$[XYZ] > \frac{1}{4}[ABC].$$

Thus, (i) is proven.

(ii)   If the positions of $X, Y, Z$ are the same as in (i), then $[AYZ]+[BZX]+[CXY] \leq \frac{3}{4}[ABC]$ implies one of $[AYZ], [BZX], [CXY]$ is not greater than $\frac{1}{4}[ABC]$, so is not greater than $[XYZ]$ by the result of (i).

If $X, Y, Z$ are not positioned as in (i), then there must be two of them, say $Z$ and $Y$ that are both above the midpoints of $AB$ and $AC$, then the distance from $X$ to the line $ZY$ must not be less than the distance from $C$ or from $B$ to the line $ZY$, so is greater than the distance from $A$ to the line $ZY$. Thus,

$$[XZY] > [AZY]$$

since they have same base $ZY$.

6.   From $a + b + c = 2$ we have $0 < a, b, c < 1$, so that

$$0 < (1-a)(1-b)(1-c) \leq \left(\frac{1-a+1-b+1-c}{3}\right)^3 = \frac{1}{27},$$

$$\therefore 0 < 1 - (a+b+c) + (ab+bc+ca) - abc \leq \frac{1}{27},$$

i.e. $0 < (ab + bc + ca) - 1 - abc \leq \frac{1}{27}$, the conclusion is proven at once.

# Index